AF887765

THE LATEST IN THE CAMPAIGN AGAINST FACEBOOK

What's Wrong with Facebook and What We Can Do to Fix It

by Gini Graham Scott, PhD, JD

THE LATEST IN THE CAMPAIGN AGAINST FACEBOOK

TABLE OF CONTENTS

PREFACE

I began writing this book after I lost my Facebook account. First, my personal account got hacked and I couldn't recover it, because I couldn't pass their selfie test to show I was who I said I was. Then, I lost my business account after I began using that and invited people to join. Facebook recommended certain people to invite after I had built up the group to about 80 members. But when I invited some of their recommendations, Facebook decided I had violated their community standards and my account was canceled with no possible appeal.

After that, I discovered many thousands of other users had similarly lost accounts due to problems with selfies and arbitrary cancellations of accounts, and I learned about the litigation against Facebook for problems with teenage addiction and privacy violations. I even learned that many advertisers not only had their ad cancelled due to supposed violations against Facebook policies, but lost the money they paid for their ads, too.

Thus, I think it's time for the government to further regulate Facebook because of its many misdeeds and continued social media power, though it has been declining, as other social media platforms gain an increasing number of users.

Here's a short overview of Facebook's decline, before I discuss what's wrong with it.

The Decline of Facebook

Facebook is still enormous, but it has been losing its position as the center of the online universe, in part because a great many people and businesses feel burned by it, and because its enforcement systems, especially selfie verification, are pushing users out of the game.

This decline has occurred over the last decade. For most of the 2010s, saying "social media" was basically the same as saying "Facebook." It was where you checked in, argued with relatives, posted vacation photos, and maybe even built your business. Today, you can feel that grip loosening.

While Facebook is still massive since it has about three billion people logging in every month around the world, other social media platforms have caught up or overtaken it in cultural importance. Instagram has grown into its own three-billion-user powerhouse, and TikTok rocketed from novelty app to global habit, especially among teens and young adults. A lot of younger users still technically "have" Facebook, but they spend their time on TikTok, Snapchat, Instagram, or in private message threads. Even older users talk about checking Facebook out of habit, not enthusiasm.

Search interest tells part of the story. Global searches for the word "Facebook" have sunk dramatically from their peak earlier in the 2020s. In the United States, surveys have found that Facebook's usage has declined among the general population over recent years. At one point, more than two-thirds of Americans 12 and older said they used Facebook. Within a year, that number slid down several percentage points — enough to wipe out a couple of years of growth in a single move. Facebook is still one of the top platforms, but for the first time the line on the chart tilted downward instead of up.

Meanwhile, platforms inside and outside of Meta's own ecosystem are surging. Instagram and WhatsApp have grown in popularity. TikTok commands an outsized share of daily attention and cultural conversation. YouTube continues to dominate long-form video. Facebook, by comparison, increasingly looks like legacy software: too entrenched to disappear, and too compromised to feel exciting. Thus, while Facebook is still a major social media player, it feels smaller. The footprint is still there; the energy is not.

How Facebook Is Losing Its Attraction for Groups and Communities

For a long time, Facebook Groups were the heart of the platform. This is where the real magic happened. There were local parenting circles, niche hobbies, professional masterminds, community organizers, and support groups on every subject you could imagine. But today, many of those rooms feel quieter.

On paper, the numbers still look impressive. Facebook has boasted that more than a billion and a half people are members of it groups. But that headline barely moved for years. Groups, like the rest of the platform, reached maturity and then seemed to stall. Membership totals may be similar, yet countless admins report the same thing: engagement is down, posts reach fewer people, and the conversations that do happen are slower and thinner.

Part of the blame lies with the algorithm. As Facebook shifted its main feed to prioritize Reels and algorithmically recommended content, Group posts started to appear less consistently, even for members who genuinely care. A teacher who used to see every update from her professional mastermind group suddenly got buried under videos from pages she never followed.

A local community group that once had dozens of comments per post now struggled to get a handful of responses.

Many users respond the way people always respond when they feel a space slipping out of their control: they leave, quietly. Or they stay but emotionally disengage. They move their core conversations to Discord servers, Slack channels, WhatsApp threads, or other platforms where the owner of the space — not some distant algorithm — still feels in charge.

Facebook Groups haven't collapsed, but they have lost something crucial: the feeling of being the default place to build community. In game terms, the lobby is still there, the player count is still high, but fewer people are logging on, because the matches no longer feel fun or fair.

How Facebook's Algorithm Has Been Kicking Out Its Users

A key problem for Facebook is that it is wrongly kicking out its users, because its algorithm has incorrectly targeted a user for misbehavior or because its selfie system to validate a person's legitimacy doesn't work. In short, Facebook feels hostile to a great many users today because of its enforcement systems. Its approach to managing its membership is like requiring its members to play a game where the referee is invisible, the rulebook is secret, and if you're ejected, there's almost no way to appeal. But if you are an active player, the game controls your social life, your reputation, and for many business owners, your income. That's what Facebook has become for countless individuals and businesses whose accounts have been disabled or locked, often with little explanation, and sometimes by mistake.

Here are some examples of individuals whose social life or business has been undermined by Facebook's policies.

- **A Photographer Who Lost Her Clients**

One victim was Maria, a portrait photographer who built her entire client base on Facebook and Instagram. For years, she ran targeted ads to reach local families, posted her work in community Groups, and managed all her bookings through Messenger and Instagram DMs. Her social profiles were her storefront, portfolio, and booking system all in one.

One morning, she woke up to a cascade of messages from confused clients. Her business page had vanished. Her personal profile was disabled. Her Instagram account, linked through the same Meta login, was gone, too. When she tried to log in, she saw a shocking notice: her account was disabled for "child sexual exploitation." But she had never posted anything remotely sexual, let alone involving minors. Her work mostly featured family portraits, graduation pictures, and headshots, in some cases of children, that had somehow triggered an algorithm for child abuse.

Even though Facebook directed her to an automated appeal process where she filled out the form, attached explanations, and waited, days later she received a generic rejection: the decision was final. But there was no human name, no contact information, and no way to prove that the algorithm had misread a perfectly innocent photo or misclassified her content.

In a single click, years of trust, connection, and business disappeared. Clients who had booked sessions through Messenger suddenly had no way to reach her. Leads gathered over years were gone. Her entire portfolio of social proof — thousands of likes, comments, and shares that reassured new customers — was erased. And besides losing access to a social network, she had been publicly labeled as the worst kind of offender by a machine she couldn't confront.

- **The Small Business Locked Behind the Glass**

Another victim was David, who ran a specialized online store selling parts for classic cars. Facebook had been his main marketing channel for years. He ran a modest but steady stream of ads that generated a reliable flow of orders each month.

Then, one day, an automated system flagged his account for "suspicious activity." He was asked to confirm his identity through a selfie verification step: he had to upload a live selfie to prove that he was a real person.

He tried to create the selfie on his phone, but the camera glitched. He tried again, but the system wouldn't accept the image. He tried on a different device; now the screen wouldn't advance past the first step. Days went by. Every attempt ended in some new error message that offered no human contact and no alternative.

In the meantime, his ad account was restricted. His active campaigns stopped serving him clients. Traffic to his website dropped overnight. When he submitted support tickets, he received canned responses telling him to complete the selfie verification flow — the same broken loop he couldn't escape. His business was essentially locked behind glass. He could see his disabled account, his saved audiences, his past campaigns, but he couldn't touch any of it. The machine had decided that the only path forward was a selfie, and when that system failed, so did much of his livelihood that depended on Facebook for clients.

- **How a Community Organizer Got Silenced**

Finally, Alisha, who ran a grassroots community group for parents advocating better safety measures in local schools, got silenced when she was wrongly accused of using hate speech. Before that happened, her Facebook Group had become a hub, through which she organized meetings, shared documents, and coordinated public comments.

One evening, after posting about a particularly sensitive local topic, she discovered that her personal account had been restricted for "hate speech." She had quoted a local official's own words and criticized them, but the automated system decided her post crossed the line. As a result, her Group lost its administrator. Scheduled events disappeared. New member requests went unanswered. Within forty-eight hours, the entire community was shut down, like a car suddenly losing its engine mid-highway.

Though Aliah tried to appeal, her appeals went into a black box. A week later, she received a confirmation that the original decision had been upheld. There was no way to reach a person, no independent review, and no trace of how the algorithm had interpreted her criticism as hate. The message was clear: speak loudly about sensitive issues on Facebook, and you might lose the very tools you're using to organize and criticize.

The Problem with Facebook's Selfie Gate: Verification Becoming a Basis for Exclusion

One of Facebook's biggest problems is the system it set up to fight bots, fraud, and harmful content -- an automated verification system based on taking a video-selfie by using your camera to submit a live image of your face to confirm your identity, so you can gain or regain access.

In theory, this verification method may sound reasonable, but in practice, it becomes a gate that many people never get through. When a selfie verification works perfectly, a legitimate user passes through a mildly uncomfortable privacy hurdle and remains a Facebook user. But when it fails because of a technical bug, a mismatch in the facial recognition system, or a misclassification, there's often nowhere else to go. The platform treats the failed selfie as a closed door and the user is excluded from Facebook.

Unfortunately, users who are already vulnerable are hit the hardest. This includes people who are less comfortable with technology, those with older devices or poor cameras, people whose appearance has changed significantly due to illness, transition, or disability, and those living in places with limited connectivity. They can all find themselves locked out by Facebook simply because the machine didn't like their image.

Another big problem with the selfie system is it often shows up on top of an already flawed enforcement decision. First, the algorithm decides that you might be a bot or a rule-breaker. Then, to "prove" you aren't, it demands an image of your face. If you're unwilling or unable to provide it, or if the system malfunctions while you take your picture, you effectively lose your digital identity on the platform.

The harm is not just an inconvenience. For active users, it is an exclusion in that they are removed from social circles, business networks, advocacy spaces, and, in some cases, from their primary source of income. The platform's refusal to provide human alternative routes or straightforward customer service turns a security feature into a trap.

The Declining Trust in Facebook

These new policies by Facebook and the resulting fallout for legitimate users have resulted in an increasing loss of trust and commitment to the platform. One irony is that Meta publicly celebrates the scale of its enforcement efforts. The company regularly reports impressive-sounding numbers about how many accounts and pieces of content it has removed for policy violations. In a single category such as sexual exploitation, it has described taking down huge numbers of accounts and posts across Facebook and Instagram.

On the one hand, those numbers are reassuring, since we want platforms to be aggressive in fighting truly harmful content. But on the other hand, they hide a crucial missing piece: the large error rates. When a system removes tens or hundreds of thousands of accounts, even a small percentage of mistakes translates into thousands of real people wrongly condemned, silenced, or erased.

From the outside, it's impossible to know how many of those removals are unfair. Meta does not routinely publish how many users eventually win their appeals or are unable to appeal at all. The stories we do hear — like Maria's, David's, and Alisha's — are likely the tip of a much larger iceberg.

This is where trust erodes. People can accept tough rules, if those rules feel consistent, legible, and appealable. What they cannot accept — at least not for long — is a system where the "referee" is invisible, the appeals process is opaque, and the consequences of one bad call are catastrophic and permanent.

When enough users and business owners have been burned like this, they start to behave differently. They post less. They rely less on Facebook for their groups and contacts with friends and family. Instead, they diversify to other platforms, mailing lists, and channels. They quietly decide, "I can't trust this place," and move their emotional energy elsewhere.

The platform's community doesn't collapse in a dramatic exodus. It thins out. The most engaged, the most creative, and the most entrepreneurial people begin to treat Facebook as a backup channel instead of their home base. So, in the last few years, Facebook has been declining as it has lost trust, and those who have lost trust, depart for other platforms.

The Irony of the Increase in Facebook's Advertising

The irony of the declining interest and trust in Facebook's platform is that it is combined with the continued growth of Meta's advertising revenue. In recent years, Meta has reported over $190 billion in annual advertising revenue, with total revenue growth in the 20-plus-percent range year over year. While the number of ad impressions has risen, the average price per ad has climbed at the same time. So the system is doing what it was designed to do: finding more ways to show more ads to more people, for more money.

While it may seem like a contradiction that people are losing faith in Facebook, as the ad dollars are increasing, from Meta's perspective, the health of the business is measured in the total revenue pulled in. So Meta doesn't care about the individual Facebook user who is wrongfully banned, or a small business whose ad account is shuttered without explanation. The company's incentives are not aligned with fixing their pain, especially if the cost of fixing it by expanding human support, building better appeals, or slowing enforcement to be more accurate, threatens the efficiency of the overall machine.

This is one reason why there is a need for more public awareness about what Facebook has been doing and for some government regulation to make Facebook, because of its huge size, more aligned with the public interest, along with making a profit. Currently, the platform is "successful" in business terms while deteriorating in human terms, in that advertising dollars can climb at the same time that trust, goodwill, and community vitality decline.

In effect, Facebook has optimized its revenue, impressions, and time on the site at the cost of the user's experience. That's why there is a need to disclose to the public what Facebook has been doing and why the government should establish some regulations to make a company with such a large public following more responsive to the needs of that public, beyond just increasing its earnings from users.

Summing Up the Need to Regulate Facebook

When you put all of these facts about Facebook and its users together, and you get a clear narrative:

- The platform is still huge, but it is no longer the cultural center of gravity.
- Groups and communities that once felt vibrant now often feel stagnant, disrupted by algorithm shifts and falling engagement.
- Automated enforcement and selfie verification systems silently eject innocent people, with little recourse and enormous real-world consequences.
- Meta measures its success in ad dollars and enforcement volume, not in fairness, transparency, or the health of its communities.

Thus, there is a need to bring back the spirit of fairness and trust to Facebook. Individual users who have been unfairly ejected from the platform may not be aware that this is a problem that has affected many thousands of users beside themselves. Individuals who have dropped out of Facebook or spend less time on it may not be aware of its growing decline with other users.

This book is designed to describe the many problems of Facebook today – from litigation to its declining appeal to users -- and discuss what can be done to fix these problems.

INTRODUCTION

This book begins with the recent litigation against Facebook, dealing with its addiction of teenage users and its privacy violations, and what the government should do about these problems. Then the book features a series of articles and press releases that were written about the problems of Facebook and what the federal and state governments should do to require Facebook to fix these problems and compensate users and former users negatively affected by these policies. As described in these articles and press releases, these problems include an inability to create a selfie resulting in the loss of an account, advertising charges due to bots and scammers, advertisers having ads cancelled and losing the money they already paid, people losing an account due to an algorithm selecting it for termination for no good reason, and more.

The articles describe these problems in detail, and the press releases feature highlights from these articles. A video highlighting these problems is at https://youtu.be/fQWD1y2rPOU.

Also, this book expands upon a previous book *What's Wrong with Facebook* that describes many of these problems.

The first section features litigation Facebook faces and how the government should respond. Then, it includes the latest article describing what the federal and state governments should do. The next section features the articles in the order in which they were written. The last section features the press releases.

THE LITIGATION AGAINST FACEBOOK

CHAPTER 1: META'S TEEN ADDICTION LAWSUITS

How Meta Ended Up on Trial for Addicting Teens

For years, Facebook's parent company Meta tried to sell a reassuring story about its products and young people: social media connects teens, helps them find support, and most kids use it just fine.

But beneath that narrative, a very different picture has taken shape in court filings, internal research leaks, and live testimony in front of juries. Thousands of families and school districts say Facebook and Instagram are not neutral tools, but deliberately engineered systems that hook children and teenagers into endless scrolling, push harmful content into their feeds, and leave a trail of anxiety, depression, self-harm, and suicide attempts.

By early 2026, that conflict boiled over into something Facebook never faced at this scale before: a coordinated legal assault that treats its youth-facing products as defective and addictive by design. Meta now sits alongside TikTok, Snapchat, and YouTube in a sprawling federal multidistrict litigation (MDL) over "social media adolescent addiction." This addiction has had serious continuing consequences. For example, in a landmark state-court trial in Los Angeles, a young woman described in open court how Instagram and YouTube consumed her childhood.

This litigation marks a major watershed for Facebook, since it is no longer just being criticized in hearings and op-eds. Now it is being asked, under oath, whether it has built systems that knowingly put children in harm's way.

Why Facebook Got Sued

The youth-addiction cases against Meta grew out of a convergence of evidence and frustration. One source of evidence was families watching their children disappear into their screens and then into emergency rooms. Many plaintiffs described a similar pattern: a child starts using Instagram or Facebook around age 10 to 13, gradually shifts to hours a day of use, and then develops depression, anxiety, eating disorders, self-harm behaviors, or suicide attempts. Parents say they watched their teens become obsessed with getting likes and followers and with the shifting hierarchies of online popularity. At the same time, Facebook's algorithms, acting behind the scenes, filled their feeds with body-comparison content, self-harm posts, or bullying.

Another source of evidence was school districts overwhelmed by a surge of mental-health crises, cyberbullying incidents, and social-media-fueled conflicts that spilled into classrooms. Districts from Maryland to Arizona claimed they had to spend millions more on counselors, crisis teams, and training to keep up with harms they traced back to social media use.

Adding to these claims was the steady drip of internal Facebook documents. In 2021, leaks and congressional materials revealed that Facebook's own research staff had identified serious risks for teens, especially girls, tied to Instagram's design. One now-famous internal slide put it bluntly: "We make body image issues worse for one in three teen girls" who struggle with those issues. Other slides said that teens "blame Instagram for increases in the rate of anxiety and depression," and some teens report experiencing suicidal thoughts when they were on Instagram.

Meta's Internal Documents Showing the Risks of Addiction

Those internal findings contradicted Meta's public posture. Outwardly, the company argued that its social media presence was largely neutral or positive, that the science was mixed, and that critics were cherry-picking extreme cases. But internally, researchers were telling executives that they had a real problem with youth harms even as externally, Facebook often minimized the risks and talked up small safety tweaks.

By 2022 and 2023, state attorneys general, personal-injury lawyers, and school-district attorneys had the basis of a case that did not just attack individual posts or failings of content moderation, but the architecture of Facebook and Instagram.

The core claim in these suits is that Meta made a conscious design choice: build platforms that maximize time-on-site, especially among young users, even if that means exploiting adolescent vulnerabilities. That means that addiction is not an unintended side effect; it is the business model for Facebook and Instagram.

Several design choices show up again and again in complaints and expert reports:

- **Infinite scroll and autoplay features that keep teens scrolling.** The feed on Facebook and Instagram is intentionally bottomless. There is no natural stopping point. As you scroll, new posts load. As you finish one video, the next auto-plays. Psychologists call this "removing stopping cues," and it is a classic way to keep behavior going longer than a person intended.

- **Algorithmic feeds that trigger the user's compulsion to remain engaged.** Rather than simply showing posts in time order, Meta's ranking systems learn what keeps each user — especially a teen — engaged and feeds them more of that content. Even if the feed is destructive, such as feeding a teen self-depreciating jokes, the algorithms can serve the user more of it, creating spirals that are hard to escape.
- **Likes, followers, and public scoreboards that provide rewards for users.** Facebook and Instagram turned social validation into visible metrics. For a teenager, every post carries the risk of public humiliation or the rush of public approval. Engagement researchers inside Meta have openly described these features as "variable reward" systems that reinforce checking and posting in the same way slot machines hook gamblers.
- **Notifications engineered to interrupt and promote more engagement.** Such notifications include push alerts, such as likes, tags, comments, "friend suggestions," and "you might like this" nudges, which pull users back into the apps repeatedly throughout the day. Plaintiffs argue these are carefully tuned to trigger curiosity and anxiety, especially among teens who already worry about missing out.
- **Weak age-gating and easy workarounds.** Although Meta says its apps are for ages 13 and up, multiple investigations have shown that younger children can join simply by entering a fake birthday. Once inside, they are treated as older teens or adults unless they are caught, meaning they can be fully exposed to the same feeds and features as everyone else.

Put together, the lawsuits say, these features form an "addictive loop" that preys on the children's and teens' developing brains. The federal MDL's own summary calls this method of entrapping users "addictive by design," noting that unpredictable likes, comments, and notifications create dopamine responses "similar to slot machines," and that teens spending more than three hours a day on social media face roughly double the risk of mental-health problems.

For Facebook, the most damning aspect of these allegations is not just that other people make these charges, but that internal teams have said similar things in more clinical language. One internal project, later dubbed "Project Mercury" in court filings, reportedly found that when users deactivated Facebook or Instagram for a week, their depression, anxiety, and loneliness scores improved — evidence that using the platforms was at least sometimes causing harm. According to those filings, Meta scaled back that research rather than drawing on it to aggressively redesign the system to result in less addictive behavior.

How Meta Tried to Fix and Spin the Problem

Eventually, though, in 2025, facing mounting criticism, Meta started rolling out more visible safety features for younger users. Instagram introduced "Teen Accounts" with stricter defaults: private accounts by default, limits on who can message or tag teens, more restrictive content filters, and parental-supervision tools linked through a "Family Center." In 2025, Meta rebranded these protections as akin to PG-13 ratings: under-18 accounts, the company said, would see content "comparable to a PG-13 movie" in Explore and recommendations.

That messaging was meant to reassure parents and regulators. But a series of independent tests undermined the story. Child-safety organizations in the U.K. and U.S. created teen test accounts with all Teen Account protections enabled and then watched what Instagram actually delivered. Their findings were stark: out of 47 safety tools Meta advertised for teens, only eight worked as described; nearly two-thirds were "substantially ineffective."

Even with PG-13 Teen Accounts, test accounts were still seeing and being steered toward the following content:

- Suicide and self-harm content, including posts that described or glamorized self-injury.
- Eating-disorder content and negative-body-image posts that encouraged unhealthy weight loss or obsession with thinness.
- Sexualized videos of minors and adult comments that were overtly sexual or predatory.
- Violent and graphic clips edited to highlight disturbing moments.

The same reports found that Instagram's recommendation systems often undermined its own filters: auto-complete suggested self-harm and eating-disorder queries, and the algorithms rewarded minors, including under-13-year-old users who had lied about their age, for posting risky, sexualized content by showering them with likes and attention.

Meta called these reports "misleading" and insisted Teen Accounts "lead the industry" in protections. But from the plaintiffs' perspective, the PG-13 branding became more evidence of deception: Facebook was promising movie-style guardrails while its algorithms continued to surface precisely the content those guardrails were supposed to block.

The Lawsuits Take Shape

By 2022, hundreds of youth-addiction suits against Meta and other platforms were dotting the federal docket. To prevent chaos, the Judicial Panel on Multidistrict Litigation consolidated them into a single case: MDL-3047, In re Social Media Adolescent Addiction/Personal Injury Products Liability Litigation, overseen by Judge Yvonne Gonzalez Rogers in the Northern District of California.

The multi-district litigation now has more than 2,200 pending cases and over 2,400 total filings, making it one of the fastest-growing MDL's in the country. About half of the cases involve Meta's products—Facebook and Instagram. The plaintiffs fall into two broad groups:

- **Individual teens and their families**, who claim that heavy use of Instagram or Facebook before age 18 caused or worsened the teens' depression, anxiety, eating disorders, self-harm, or suicidal ideation.
- **School districts**, which say the social-media addiction of their students forced them to spend millions on mental-health staff, threat-assessment teams, and classroom management resources, and to cope with lost instructional time and constant crises.

Judge Gonzalez Rogers has allowed a large portion of these claims to proceed past Meta's attempts to throw them out. She has drawn an important line: Section 230 may protect platforms from being treated as the publisher of user content, but it does not automatically protect the design of the product itself. Claims focused on infinite scroll, notification design, weak age-gating, and failure to warn parents can go forward even where content-based claims cannot.

Parallel to the multi-district litigation, more than 40 state attorneys general, from California and New York to smaller states, have sued Meta under state consumer-protection and unfair-practice laws. Their complaints lean heavily on internal research, alleging that Meta marketed Instagram as safe for teens while its own data showed serious mental-health risks, and that it misrepresented or obscured how its recommendation systems actually worked.

The Landmark Los Angeles Trial

At the same time, one individual case has leapt ahead of other MDL cases and become a full state-court trial: Kaley G.M. v. Meta Platforms, Inc. and Google LLC, in Los Angeles County.

Kaley is a 20-year-old woman from California who began using YouTube at six-years-old and Instagram at age nine. She says that as she grew up, her use of these platforms turned compulsive: she engaged in endless scrolling late at night, posting and checking constantly, and measuring her worth through likes and comments. According to her testimony and expert witnesses, this compulsive use coincided with the onset of depression, body dysmorphia, self-harm, and suicidal thoughts.

Her lawyers argue that Meta and YouTube deliberately designed features — auto-scroll, autoplay, like systems, infinite feeds — to hook her, and that they did so knowing from internal research that such designs were especially risky for children. The trial has already featured testimony from Kaley's former therapist, who told jurors that her excessive social-media use was "a significant contributor" to her mental-health struggles.

Meta and YouTube have pushed back hard. Both companies reject the idea that social media is inherently "addictive," and they say that the evidence does not support Kaley's specific claims. Meta has told reporters and jurors that Kaley "faced many significant, difficult challenges well before she ever used social media," and that the question is whether Instagram was truly a substantial factor in her injuries. Both companies point to billions of dollars spent on safety features, including teen-specific accounts and parental controls, and they argue that many of those tools simply did not exist when Kaley was a child.

Mark Zuckerberg himself took the stand in February 2026, defending Meta's choices and insisting that the company never set out to harm children. But the trial, which continues as of late February 2026, has already marked a turning point: for the first time, a jury is hearing about how Facebook's design decisions might have ruined one young person's life.

Where the MDL Cases Stand Now

While Kaley's case plays out in Los Angeles, the multi-district litigation in Oakland is grinding forward. As of early 2026, more than 2,325 social-media addiction claims are pending there, with new filings continuing to arrive.

Judge Gonzalez Rogers has selected a set of bellwether cases to test in federal court. Among them are school districts from Maryland, Georgia, Kentucky, New Jersey, South Carolina, and Arizona that will be the first to present their claims to a jury. One of those districts — Harford County Public Schools in Maryland — has publicly estimated more than $41 million in past damages and up to $1.2 billion in future abatement costs tied to social-media-related harms among its nearly 40,000 students.

Those bellwether trials are expected to begin later in 2026. Their outcomes will heavily influence whether Meta and other defendants negotiate a global settlement, fight cases one by one, or some mix of both. In the meantime, Meta finds itself trying to hold two lines at once:

- In court, it argues that Section 230 and the First Amendment limit its liability, that plaintiffs cannot prove causation, and that it acted reasonably by adding safety features over time.
- In public, it continues to roll out new teen protections — like alerts to parents if a teen repeatedly searches for self-harm content, while it emphasizes its "longstanding commitment to supporting young people."

Critics see those new features as too little, too late, and, in some cases, as reactive moves designed to blunt regulatory and legal pressure rather than to solve the underlying problems.

The Latest Developments

Two threads define the state of play in early 2026: the evolution of Meta's safety features, and the momentum of its legal cases.

On the product side, Meta is trying to show that it is evolving. Beyond Teen Accounts and PG-13 filters, the company has:

- Expanded teen protections to Facebook and Messenger, not just Instagram, including requiring parental permission for younger teens to go live and tightening how minors are surfaced in recommendations.
- Announced new alerts that notify parents when a Teen Account repeatedly searches for suicide or self-harm content, positioning this as an added layer on top of hiding or de-ranking such content in feeds.
- Continued to tweak content controls and messaging rules, insisting that adults "cannot send DM requests to teens who don't follow them" and that teens are "recommended less potentially sensitive content."

At the same time, new reporting keeps undercutting Meta's claims. Investigations into internal emails show growth teams worrying that making teen accounts private by default would cost millions of monthly active users, even as safety teams warned that public profiles were driving millions of unwanted adult–teen interactions per day. External tests keep finding under-13s on the platform and harmful content slipping past PG-13 filters.

On the legal side, the first wave of trials is already reshaping expectations. In Kaley's case, two co-defendants—Snap and TikTok—chose to settle before trial, leaving Meta and YouTube to face the jury alone. Even without knowing the exact settlement terms, that decision signals that some platforms believe juries may be sympathetic to the "addicted by design" theory and willing to award significant damages.

And hovering over everything is the U.S. Surgeon General's warning that "we cannot conclude social media is sufficiently safe for children and adolescents," noting that adolescents who spend more than three hours per day on social media face about twice the risk of mental-health problems and that nearly half of teens say social media makes them feel worse about their bodies. That advisory decision gives judges, juries, and legislators a simple, authoritative frame: whatever Facebook and other platforms have done so far, it has not been enough.

What's Really Wrong with Facebook

This youth-addiction litigation crystallizes many of Meta's platforms' deeper flaws.

First, it exposes a familiar pattern: internal knowledge versus external denial. Facebook's own researchers identified serious harms to teen mental health and body image; its executives saw the slides; yet the company continued to optimize for engagement, slow-walked structural fixes like default-private teen accounts, and downplayed risks in public. That is the same playbook the company followed on privacy, misinformation, and political polarization.

Second, this litigation shows how design choices are moral choices in that the infinite scroll, algorithmic feeds, and badge-like notifications are not neutral; they are tuned systems built to maximize time and ad impressions. For adults, that may be annoying or distracting. For children and teenagers, those same choices can be destabilizing or even deadly. The lawsuits force Facebook to defend not just what appears in the feed, but the feed itself.

Third, this litigation underlines that Facebook has systematically treated minors as growth assets, while internal strategies at Instagram have explicitly centered on "winning back teens" from TikTok and YouTube, even as Meta was under investigation for youth harms. The company built "teen museums" in its offices to help staff think more like adolescents and turned teen engagement into a key success metric. Teen-branded safety tools like Teen Accounts have to be read against that backdrop: are they genuine guardrails, or are they PR armor over a youth-growth engine?

Finally, the youth-addiction cases reveal how difficult it is to rein in a company structured like Facebook through voluntary measures. For years, Meta has promised to do better, launched partial fixes, and issued blog posts about "well-being." Yet external audits keep finding that the underlying incentives to keep teens scrolling and keep ads flowing still win. That is why so many plaintiffs, regulators, and advocates are now looking to the courts and regulators to do what Facebook has not done for itself: put hard, enforceable limits on how it can treat children.

Though Facebook is not the only platform faced with litigation, since TikTok, YouTube, and Snapchat are being sued alongside Meta and share many of the same addictive design patterns, Facebook has a central role in these cases. That's because of its long trail of internal research, its central role in the multi-district litigation, and its position at the heart of the Kaley trial make it the most important defendant in these cases. If juries and judges decide that Facebook's youth-facing products are defective by design, it will not just reshape one company. It will redraw the line between acceptable and unacceptable behavior for every platform that seeks to profit from the attention of children.

The Seriousness of Social Media Addiction

This legal effort to go after Meta and the other platforms causing social media addiction is supported by numerous experts.

For example, in the Los Angeles trial against Meta and YouTube, one expert described that these apps work like a "A slot machine that fits in your pocket", where you pull a lever and get an unpredictable reward, such as a flattering comment, an exciting video, or nothing at all. In turn, that unpredictability drives people to keep pulling the lever or swiping the feed, and rather than money, their payout is a dopamine hit in the brain which occurs when we see something pleasurable or socially rewarding. Then that hit reinforces the behavior and makes it more likely we'll repeat it.

Another expert, Stanford psychiatrist Anna Lembke, who has studied digital addiction, compared the effect of these reward loops to drugs. Early use brings pleasure through positive reinforcement, but as habits take hold, users return to the app mainly to stop feeling bad by relieving their anxiety, boredom, or withdrawal experiences if they try to reduce or stop using the app. That shift from "this feels good" to "I feel awful without it" is a hallmark of addiction.

Even Facebook researchers back in 2018 became worried about the way its "design elements…might encourage compulsive usage." As the researchers noted, certain features could "encourage frequent, automatic, unwanted behaviors" and lead to usage patterns that left people feeling "manipulated," "lacking control," and even "addicted" to Facebook and Instagram. However, while the researchers suggested partnering with outside experts to study these effects, according to internal emails, the Facebook product teams feared such a collaboration would lead to pressure for design changes they did not want to make. That tension between safety researchers wanting to test for harm and product teams worried about growth is exactly what plaintiffs now highlight in court.

How the Trials May Lead to Certain Reforms

Whatever happens in the litigation against Meta, the expert testimony and multi-district litigation are already shaping the policy conversation, resulting in certain types of changes if the plaintiffs win meaningful verdicts. These include the following changes that Meta will be required to make:

- Establish stronger design rules for youth accounts, resulting in hard limits on certain features for those under 18, such as stricter caps on notifications and limits on infinite scrolling.
- Create mandatory default safety settings, not just optional ones, such as setting up strict content filters.

- Setting up broader youth-data firewalls in order to require platforms to segregate and heavily restrict how they can use minors' data for advertising or recommendations.
- Provide more algorithmic transparency and audits, so that companies must let regulators and vetted researchers know about the potential youth exposure to harmful content.
- Institute regular, independent safety audits of teens using different features of the platform, along with public summaries and giving the government the power to compel fixes.

.

What Reforms *Should* Happen

Regardless of what happens in the current lawsuits against Meta, its Facebook and Instagram platforms should be required to do the following:

Change Incentives

Based on research showing the problem of addiction to social media and Facebook's own internal research, the most meaningful reforms are those that change incentives and design, not just labels and settings. Here are some reforms that should happen.

- Facebook should be legally required to behave less like an advertising platform and more like a regulated health product.
- For users under-18, feeds should provide a chronological or simple "friends and groups" sorting, not engagement-optimized ranking that amplifies outrage, comparison, or shock.
- Facebook should be prevented from using features that provide dopamine hits, such as an infinite scroll, autoplay, and unpredictable reward notifications.

Such requirements would force Meta, as well as other social media platforms, to compete on usefulness and genuine connection not on who can create the strongest habit loops in vulnerable users.

Require Safety by Default

Since voluntary teen protections have repeatedly proven weaker in practice, the following safety practices should be required.

- The strictest safety settings should be required by default for all users under 18, with no ability to turn those protections off. Parents could choose to add additional restrictions, but not to drop below the required minimum of safety settings.
- If a device is known to be used by a minor, every account on it should be treated as a minor's account unless proven otherwise, closing the "secret adult account" loophole.

Require the Findings of Internal Research to Shape External Behavior

Since there has been a gap between what its scientists find and what Meta's products do, regulators should require Meta and other social media platforms to do the following:

- Require large platforms to maintain an independent safety research function with a duty to report serious risks to regulators.

- Mandate that any internal research identifying significant youth harms be disclosed to the FTC and relevant state Attorney Generals, with a timeline for proposing mitigations.
- Make it an unfair or deceptive practice to publicly downplay substantial harms that the company's internal science identifies, exposing companies to fines and structural remedies if they don't make the necessary changes to avoid the harms.

Such requirements would make it much harder for Facebook, Instagram, or other social media platforms to bury or spin results that show the addiction of teens.

Build Independent Oversight with Real Teeth

Since voluntary corporate promises and internal safety teams are insufficient when they interfere with the company's profits, the incentives should be changed to undermine practices that are profitable but promote addiction. These measures might include these:

- Congress or state legislatures could authorize an independent digital-safety regulator or expand the FTC's youth-safety oversight, giving them the authority to audit algorithms, mandate changes, and levy substantial penalties for non-compliance.
- Courts in the multi-district litigation could require ongoing company monitoring, such as implemented after the tobacco and opioid settlements, which would require Meta and other social media companies to fund and cooperate with long-term independent evaluations of its youth-oriented products.

The goal of such measures is not just to punish past behavior but to install a permanent counterweight to the growth-at-all-costs culture inside companies like Meta.

Pair Restrictions with Better Digital-Mental-Health Support

Finally, reforms should not only restrict harmful design but also require Meta and other social media companies to harness technology for good. Some of the ways to do so are these:

- Require large platforms with substantial youth user bases to offer evidence-based mental-health tools, such as in-app prompts that connect teens who repeatedly search for self-harm content to crisis hotlines or peer support. Plus these platforms can offer moderated spaces for mental-health education.
- Fund independent, youth-friendly mental-health services, which are accessible via the social media but are governed by healthcare and privacy rules.

What Other Countries Are Doing to Support Youth Social Media Safety

These listed reforms recommend what other countries are already doing to support youth social-media safety. These countries are treating youth safety not just as a regulatory issue or a matter for corporate promises, but they are passing legislation that impacts the social media platforms. These models hint at both the legal impacts and the financial stakes that Meta faces if it loses these cases. Here are some of the things that other countries are doing:

CHAPTER 2: META'S PRIVACY VIOLATIONS

Facebook and Meta have been at the center of some of the largest privacy scandals and lawsuits in modern history, revealing a business model built on pervasive surveillance and aggressive data monetization. This chapter looks at what they actually did, how plaintiffs and experts have framed those actions as privacy violations, how Meta defends itself, where the key cases stand today, what the U.S. government should do.

How Facebook Invades Privacy

From the beginning of its founding, Facebook's core business has depended on collecting as much personal data as possible, correlating it, and selling finely targeted access to users' attention to advertisers and political actors. Litigation and regulatory actions have forced previously hidden practices into the open, showing a pattern rather than a one-off mistake.
Key privacy-invasive behaviors include:

- **Over-collection and over-sharing of data**. Facebook allowed a wide range of app developers, data brokers, and political consulting firms to access user data far beyond what a reasonable person would expect, often including information about a user's friends who never directly consented to the sharing. At one time, in the Cambridge Analytica era to be described, a single quiz app could harvest data from tens of millions of users and their networks, feeding psychographic profiling for political ads.

- **Shadow profiles and inferred data.** Even when users didn't explicitly provide certain information, Facebook inferred it from their behavior, connections, and technical signals like their IP address. As a result, Face was able to build detailed "shadow profiles" that powered personal targeting and experimentation. Litigation over location tracking, for example, alleges that Meta continued to infer location from IP addresses and other signals even when users thought they had turned their location sharing "off."
- **Tracking beyond Facebook through pixels and SDKs.** Meta uses tracking pixels, social plugins, and mobile SDKs, software development kits, a collection of tools and guides that help developers build applications, so the company can follow users' browsing and in-app activity far outside Facebook and Instagram. In healthcare, class actions allege that Meta embedded a pixel on hospital websites and patient portals which captured sensitive health information, such as appointment details, portal logins, and communications with providers. Then, the pixel sent that information back to Meta for advertising and profiling.
- **Biometric surveillance through facial recognition.** For years, Facebook's "tag suggestions" feature quietly scanned faces in uploaded photos, extracted biometric identifiers to create facial geometry, and built a database that could recognize individuals, often without clear, informed consent, including in jurisdictions like Illinois that require explicit opt-in. That program became the core of a wave of biometric privacy litigation that was filed under Illinois' Biometric Information Privacy Act (BIPA).
- **Opaque consent and dark patterns.** Meta's interface has repeatedly buried key privacy decisions in long policies and confusing settings, while nudging users toward making disclosures that benefit advertisers. In multiple cases, Meta has argued that users "consented" to the practices at issue simply by using the platform under the terms of service and data policies, even when those terms were difficult to understand or practically impossible to avoid. And typically don't read these terms; they just accept them by using the platform.

These different types of behaviors indicate that Meta has not engaged in a single breach on Facebook but created an architecture of surveillance. While users engaged in the product to make social connection, behind the scenes, Facebook's business is behavioral data extraction at a massive scale.

The Cambridge Analytica Class Action and Other Cases Against Facebook

The central privacy case against Facebook in the United States is the sprawling multidistrict litigation known as *In re Facebook, Inc. Consumer Privacy User Profile Litigation*, which grew out of the Cambridge Analytica scandal.

This case got its start because Cambridge Analytica, a political consulting firm, obtained personal data on tens of millions of Facebook users through a personality-quiz app, which obtained this information through Facebook's Open Graph API. The app collected the data of users who installed it and additionally obtained data about their Facebook friends who had never installed the app or knowingly agreed to share their information for political profiling.

In pursuing the litigation, the plaintiffs alleged three main facts. They charged that Facebook allowed apps and third parties to have broad access to user content and information without proper consent by obtaining it from likes, posts, friend lists, and other profile data. They claimed that Facebook failed to adequately monitor or restrict how those third parties used the data, including using it for targeted political advertising and influence operations. Additionally, they alleged that Facebook misrepresented the extent of its data-sharing and the control that users had over their information.

The Plaintiffs, Experts, and Class Members in the Lawsuits

The plaintiffs in the Cambridge Analytics suit were a massive class of U.S. Facebook users whose data had been shared or exposed to third parties without their adequate consent. The law firms representing the plaintiffs included Keller Rohrback L.L.P. and Bleichmar Fonti and Auld LLP, which served as co-lead counsel and ultimately negotiated the settlement.

The plaintiffs' case relied on three main types of experts. These included technical experts who demonstrated how the Open Graph API and Facebook's platform policies actually worked in practice by showing how broad the data flows were and how little oversight existed. They also called on privacy and consumer-protection experts who explained the mismatch between Facebook's public promises and its internal practices, and they described the harms of large-scale, non-consensual data harvesting. Additionally, they called on economists who estimated the value of the data collected and the harm to users' privacy interests, supporting large aggregate damage figures.

Then, too, the litigation relied heavily on testimony from experts who explained that Facebook's practices were not "industry standard" accidents but choices designed to maximize growth and engagement while externalizing privacy risks.

Meta's Defense

In the face of these arguments, Meta consistently denied wrongdoing. It argued that the users consented to data sharing through Facebook's terms, policies, and settings, and could control what third-party apps accessed. It claimed there was no concrete harm to most users, making large-scale damages inappropriate.

Also, after the scandal erupted, the company claimed it made extensive changes to limit third-party access and improve privacy controls, indicating good-faith efforts rather than malicious intent.

Then, after the company finally engaged in a settlement, in its public statements, Meta emphasized that it had "revamped its approach to privacy" and entered into the deal because it was in the best interests of users and shareholders, not as an admission of liability.

The Lawsuit Settlement

Eventually, the Cambridge Analytica litigation ended with a historic settlement that provided for the following terms:

- In 2022 Meta agreed to pay $725 million to resolve the case, the largest privacy class-action settlement in U.S. history.
- In October 2023, U.S. District Judge Vince Chhabria granted final approval to the settlement in the Northern District of California.
- Between 250 million and 280 million U.S. Facebook users were considered eligible for compensation, with individual payouts depending on how many filed valid claims.

However, since the parties agreed to settle, A U.S. appeals court has heard objections from class members who argued that the settlement amount was too low and that attorneys' fees (about $181 million) were excessive, but Meta has defended the settlement as fair and consistent with privacy-law risks.

Besides entering into the settlement, Facebook also paid a $5 billion civil penalty to the Federal Trade Commission in 2019 to settle allegations that it violated an earlier FTC order by deceiving users about how it shared their data with third-party apps, including Cambridge Analytica. That FTC case is formally about consumer protection and order violations, but it is fundamentally grounded in the same privacy failures.

Biometric Privacy: Facebook's Facial Recognition and BIPA

Besides the Cambridge class action case, Facebook also faced a challenge under the Illinois' Biometric Information Privacy Act (BIPA) which challenged Facebook's use of face recognition. It did so, since BIPA requires private entities to obtain informed, written consent before collecting biometric identifiers like facial geometry, and it provides a private right of action with statutory damages.

But for years, Facebook violated these requirements, since its "tag suggestions" feature scanned users' photos, detected faces, and created templates that could be used to recognize individuals across images. Accordingly, in the case, the plaintiffs argued that Facebook collected, stored, and used biometric identifiers of Illinois residents without obtaining the kind of explicit, written consent BIPA demands. They claimed that users were not clearly told that face-template data was being created and retained, nor were they told how long that data would be stored or for what purposes. Additionally, they charged that the system enabled ongoing, pervasive biometric surveillance in that people's faces became perpetual identifiers in Facebook's database.

The Plaintiffs and Experts

The original BIPA class actions against Facebook were brought on behalf of Illinois Facebook users whose faces appeared in photos on the platform. Plaintiffs' counsel and experts focused on three key accusations: the sensitivity and permanence of biometric identifiers, since unlike passwords, you cannot change your face if it's compromised or misused; the risk of identity theft, fraud, and broader surveillance if such facial data is repurposed or breached' and the way Facebook's system repeatedly captured and processed face data, multiplying the potential violations and damages.

Meta's Defense

In response to these charges, Meta argued that its terms of service and settings provided sufficient notice and consent, since users could opt out, that California privacy law should govern disputes because Meta is headquartered there and its terms specify California law, which would reduce or eliminate its exposure in Illinois; and that the number of actionable "violations" should be limited, because counting every scan or separate interaction as a separate violation would produce enormous, disproportionate damages. In response to this flood of BIPA litigation, Illinois lawmakers amended the statute in 2024 to clarify that multiple captures of the same identifier from the same person would count as a single violation in figuring per-person damages.

Outcomes and Status

Ultimately, Facebook reached a class-action settlement over its facial recognition practices, paying hundreds of millions of dollars to Illinois users, and eventually it retired the tag-suggestions system. However, with BIPA amended to treat repeated scans as a single violation, Meta's potential per-user exposure is reduced, but it still could be liable under Illinois law for future violations, and the courts are signaling that biometric consent cannot be contracted away.

In fact, there is a more recent case, *Hartman v. Meta Platforms, Inc.*, in the Southern District of Illinois, in which the plaintiffs are again pursuing BIPA claims, and a February 2026 ruling rejected Meta's attempt to apply California law instead of Illinois law. The court held that enforcing Meta's choice-of-law clause would undermine Illinois' fundamental public policy of protecting biometric privacy and deprive residents of BIPA's protections. So Facebook is still fighting this ongoing case there.

Still Other Major Privacy Cases and Penalties Against Meta

Beyond Pixel and Cambridge Analytica, Meta has been hit with still other privacy-related actions and has settled most of them. A few recent cases of note are these.

Biometrics in the Texas Attorney General Case

In this case, Texas sued Meta for unlawfully capturing Texans' biometric identifiers via Facebook's "tag suggestions" facial-recognition feature without informed consent. The Texas attorney general alleged that Meta violated Texas' Capture or Use of Biometric Identifier Act and the Deceptive Trade Practices Act. To settle the case, in 2024, Meta agreed to pay $1.4 billion dollars to Texas, the largest state privacy settlement to date.

EU and Irish Fines for Privacy Violations

In this case, Ireland's Data Protection Commission (DPC), acting under the EU's, General Data Protection Regulation (GDPR), its comprehensive data privacy law, effective since May 25, 2018, fined Meta multiple times. According to this law, there are strict rules on how organizations collect, use, and protect the personal data of individuals in the EU and the European Economic Area (EEA). The law gives users more control over their data, including the right to be forgotten through erasure. Based on this law, Ireland imposed multiple fines on Meta totaling hundreds of millions of euros for issues including unlawful reliance on "contract" as a legal basis for behavioral advertising and mishandling user password. The country also fined Meta a 91-million-euro fine - about $102 million in U.S. dollars - for storing some user passwords in plaintext.

Scam-Ad and Endorsement Misuse Litigation

A 2026 class action alleges Meta allowed scam investment ads that unlawfully used financial professionals' names, images, voices, and likenesses in ways that misled users and damaged their reputations, highlighting Meta's role in enabling misuse of personal identity data in ads.

A related class action lawsuit was filed in a California federal court by a group of consumers who claim that Meta enabled, facilitated, and materially contributed to a multimillion-dollar stock manipulation scheme that extracted millions of dollars from unsuspecting victims. The plaintiffs Anthony Irving, Yaakov Strauss, and Santosh Kumar filed the class action complaint against Meta on February 5, 2026, in California federal court, alleging violations of state and federal laws whereby Meta played a pivotal role in a stock manipulation scheme that used scam ads on Facebook, Instagram, and WhatsApp to defraud victims of millions of dollars. According to the plaintiffs, Meta used its advertising tools, which generate and target ads based on user data to get the ads, and then the scammers exploited these tools to create and disseminate fraudulent investment ads. In using these tools, the scammers targeted specific user groups with their ads that lured victims into a pump-and-dump scheme involving Jayud Global Logistics Ltd. (JYD), a Chinese stock listed on NASDAQ, leading the victims to purchase JYD shares at inflated prices. Then, the scammers sold their holdings at a profit before the stock's value plummeted, leaving victims with substantial losses.

According to the plaintiffs, Meta was long aware of the scam ads on its platforms but failed to implement effective measures to monitor, identify, and remove them. Instead, Meta used generative AI tools that exacerbated the problem by creating multiple variations of the scam ads, increasing their reach and effectiveness. Allegedly, Meta's failure to act to fix the problem was driven by its reliance on advertising revenue, a significant portion of which is derived from scammers. Despite promising users that it would not allow scam ads and would take action when alerted to them, Meta allegedly turned a blind eye to the problem, according to the lawsuit.

Currently, the litigants are seeking additional class members drawn into investing between March 21, 2025, and April 2, 2025, due to the fraudulent Facebook and Instagram ads and suffered losses as a result.

In still another Meta class action filed in California federal court, two plaintiffs, John Suddeth and Sara Perkins, accuse Meta of unlawfully using their financial professionals' names, images, voices, and personas in paid ads and related promotional content disseminated on Facebook, Instagram and/or WhatsApp. According to the lawsuit, the plaintiffs argue that Meta's conduct induced users to invest in fraudulent, thinly traded China-based securities. They also claim that Meta failed to act after a bipartisan coalition of state attorney generals warned them that paid impersonation ads and WhatsApp investment groups were being used to perpetrate widespread fraud against U.S. consumers. Instead, starting in July 2025, Meta continued to allow these impersonation ads and funnels to proliferate. So now the plaintiffs are seeking other U.S. financial professionals whose names, images, voices, likenesses, credentials, branding or professional personas were used without their consent in paid ads or promotional content to promote securities or investment opportunities from January 1, 2023, to the present.

Additional Cases Involving the Meta Pixel, Location, and Health Data

There are still more privacy cases, since Meta has faced other cases involving tracking individual locations and obtaining and sharing health data.

<u>Location-Tracking Class Actions</u>

Meta was involved in a series of class action cases in which it was charged with using location tracking, and it resolved those claims with a $37.5 million settlement in 2024. In these cases, the plaintiffs claimed that Meta improperly inferred the location of U.S. Facebook users from their IP addresses despite its representations that users could control location tracking. Instead, Meta continued to derive the approximate location of users for ad targeting and analytics even after users disabled precise location services or believed they had opted out. Also, the plaintiffs claimed that Meta misled users about the extent of tracking and failed to obtain meaningful consent for its IP-based location inferring where they were.

While Meta denied wrongdoing, it agreed to the settlement covering a nationwide class of users whose location data was allegedly collected between 2015 and 2022. As in other privacy cases, Meta emphasized publicly that the settlement was a business decision, not an admission of guilt.

<u>Meta Pixel in Healthcare</u>

Another especially disturbing privacy case involves Meta's Pixel tool providing confidential information to health-care providers. In this class action case in the Northern District of California, *In re Meta Pixel Healthcare Litigation,* the plaintiffs have alleged that:

- Hospitals and medical systems installed the Meta Pixel on their websites and patient portals, causing communications about appointments, portal logins, and other interactions to be sent directly to Meta. It did so in violation of the federal Electronic Communications Privacy Act (ECPA) and state wiretap statutes.
- These transmissions included protected patient health information that patients reasonably believed would remain confidential. Accordingly, the company knowingly received protected health information (PHI) from hundreds of providers without HIPAA-compliant authorization or any meaningful patient consent and then monetized that data via targeted advertising.
- Meta used the intercepted health-related browsing and portal activity for analytics and ad targeting on Facebook and Instagram, without patients' consent. As a result, Meta's own contracts and public privacy promises to users were breached, because the patients did not authorize this sort of medical surveillance

This was a particular egregious violation since Meta's tracking machinery followed people into the most private digital spaces they visit: hospital websites and patient portals.

But instead, the California patients allege that this Meta Pixel tool was quietly embedded on hundreds of hospital systems' websites and portals. As a result, when patients searched for a doctor, clicked on pages describing conditions like cancer or heart disease, registered for or logged in to a patient portal, or scheduled or managed appointments, the Pixel allegedly captured those interactions. Then, it sent them, in real time, to Meta's servers for "analytics" and ad targeting.

The plaintiffs in this case are patients whose interactions with specific hospital systems were allegedly captured by the Pixel. They are represented by a coalition of plaintiffs' firms, including Cohen Milstein Sellers & Toll, which serves as the interim co-lead class counsel. The case also calls on technical and privacy experts to reconstruct the traffic to the hospitals' websites, showing how the Pixel code works to capture this patient information and exactly what fields and URL paths are transmitted to Meta.

In one example of how this process works, a lead plaintiff noticed that after visiting a health-system portal and viewing pages about heart and blood-pressure issues, she began seeing a wave of high-blood-pressure medication ads on Facebook. This occurred, as the experts showed, because the Pixel can capture page titles, URL paths, button clicks, and form entries. Then, when this information is correlated with IP address and browser attributes, that links the health activity to a specific Facebook account.

In its defense, Meta made several arguments. It argued that HIPAA does not directly apply to Meta as a "business associate," because it has no HIPAA-style contractual relationship with providers, and HIPAA has no private right of action for patients. It claimed that any information it received was covered by its general data policies and that any collected data is filtered and aggregated, not used to identify individuals in a way that violates wiretap or consumer-protection laws. Also, it contended that the presence of Pixel on "public webpages" undermines any expectation of secrecy that would support wiretap-style claims.

So far, courts have been skeptical of Meta's attempt to make the case go away. In September 2023 and again in January 2024, Judge William Orrick denied Meta's motions to dismiss the plaintiff's core claims about privacy violations, allowing the plaintiff's claims to move forward. He rejected the argument that the use of public web pages destroys any privacy interest in patient-related communications, and he noted that Meta's own systems recognize that Pixel traffic can reveal "patient status," which counts as a privacy violation under HIPAA.

The net effect of the judge's decision at this stage is that the plaintiffs' theory of "wiretapping the doctor's office" is legally plausible, and Meta faces both damages exposure and the risk of a court order forcing structural changes in how its Pixel handles health-related data.

At the same time, separate, related suits have targeted the health systems themselves for enabling Pixel-based tracking on patient portals, leading to settlements by organizations like Catholic Health System and Northwell Health, though patients aren't allowed to sue directly through HIPPA. However, while those suits don't directly target Meta, those cases show that Meta's tools are so deeply embedded in the web that privacy exposure spreads outward to any institution that plugs them in.

How Federal Regulation Has Affected Facebook and Meta

While class actions focus on damages for users, regulatory enforcement has come primarily from the Federal Trade Commission (FTC), which has pursued Facebook for more than a decade over broken privacy promises.

The story starts with a 2012 FTC order that resolved allegations that Facebook misled users about how their information would be shared, including making data public that users had set to "friends only." That order required Facebook to stop misrepresenting the privacy or security of users' personal information, to obtain affirmative express consent before overriding privacy settings, and to implement a comprehensive privacy program subject to regular audits.

When the Cambridge Analytica scandal erupted, the FTC viewed it not just as a new privacy debacle but as evidence that Facebook had violated its 2012 order. The Commission alleged that Facebook deceived users about how their data would be shared with third-party apps and "friends' apps," failed to maintain a reasonable privacy program, and failed to tell users about certain unauthorized data uses and breaches affecting tens of millions.

As a result of these charges, in 2019, the FTC announced a record-breaking $5 billion settlement with Facebook, along with sweeping modifications to the 2012 order. These modifications now required Facebook to make the following key changes It had to set up a new privacy structure at the board level, including an independent privacy committee that the CEO cannot control. It had to designate compliance officers personally responsible for the privacy program, who, along with the CEO, had to provide quarterly certifications and annual compliance certifications to the FTC. Meta also had to expanded its audits and documentation requirements, including being obligated to report incidents in which the data of 500 or more users was compromised. Additionally, Meta was prohibited from engaging in certain practices, such as asking users for email passwords to other services as a way to obtain this contact information.

The FTC publicly framed this 2019 order as a forced "restructuring" of Facebook's approach to privacy, with individual executives on the hook for false certifications. Plus Meta was still liable for the $5 billion penalty.

Ironically, Meta has claimed that this settlement resolved past allegations and created a rigorous privacy compliance environment with new governance mechanisms, internal privacy review processes, and increased user controls as evidence that it now takes privacy seriously. Yet the continued litigation against it for privacy violations shows that Meta is still using tools like the Meta Pixel to gain individual information and use it for advertising.

Also, the relationship with the FTC has remained contentious. For example, in 2023, the FTC proposed tightening the order further, alleging that Meta still had not fully complied and seeking, among other things, to bar Meta from monetizing data collected from users under 18. Meta has fought those proposed modifications in court, casting them as regulatory overreach and arguing that the FTC is effectively trying to regulate business models without legislative authority. As of early 2026, that battle over the scope of the FTC's power and the future of Meta's data-driven advertising model is ongoing.

Summing Up Metta's Privacy Cases

Across these many privacy cases, the plaintiffs fall into several overlapping groups, each highlighting a different aspect of Meta's intrusion into personal life. To sum up, these include the following:

- **Everyday Facebook users** whose data was shared with third parties like Cambridge Analytica or other app developers without their consent; they formed the core of the $725 million Cambridge Analytica settlement class.
- **Illinois residents** whose faceprints were captured by Facebook's facial-recognition systems without the explicit written consent required by BIPA.
- **Users whose approximate geolocation was inferred** via their IP address and used for targeting ads to them, a violation of Meta's promises about location controls.
- **Patients who interacted with healthcare providers,** which resulted in their interactions being intercepted by Meta Pixel, exposing their sensitive medical information to Meta's data systems.
- **Parents and children in separate youth-harm litigation** who alleged that Meta's design choices and recommendation algorithms harmed minors; and in some cases this litigation was combined with questions about what data is collected and used.

In response, Meta sought to defend itself with an army of major law firms, in-house counsel, and technical experts who argue that Meta's disclosures and consent flows are adequate under existing law; that many alleged harms are too speculative, diffuse, or non-economic to support massive damages; that the company's current practices are in line with, or better than, industry norms; and that it has invested heavily in privacy after learning from past mistakes.

In the BIPA and Pixel cases, Meta also invoked jurisdictional and conflict-of-laws arguments, whereby it tried to keep disputes in California and under California law, where there is no BIPA-style private right of action for biometrics and the damages exposure is lower.

Defeating Meta's Claims that "We've Changed" vs. "Trust Us"

Whenever Meta settles a case or responds to a new complaint, its public narrative revolves around two claims: that the events are in the past and that the company has fundamentally changed how it treats privacy, though, in fact, Meta continues to act to support its advertising and continue its problem practices.

In support of these claims Meta highlights new privacy controls and simplified settings for users, investments in end-to-end encryption for messaging, reduced default data retention for some products, and restrictions on what third-party apps and business partners can access. To this end, in regulatory filings and court arguments, Meta has stressed that it has made the reforms imposed by the FTC order and made voluntary changes like shutting down facial recognition, limiting app permissions, and introducing new data-minimization measures. The message is that Meta has addressed its past problems and continuing to punish the company for those eras is unfair and counterproductive.

However, plaintiffs and regulators have disputed those claims by pointing out that each new scandal reveals a consistent pattern of over-collection, opaque sharing, and only retroactive reform when caught. From this perspective, the plaintiffs and regulators have pointed out these three key points:

- Cambridge Analytica was not an isolated fluke but the predictable outcome of an open-API growth strategy with weak enforcement.
- Facial recognition, Pixel-based health tracking, and IP-based location inference all use different technical methods to achieve the same strategic goal: turning human behavior into data points that can be predicted, segmented, and sold.
- "Reforms" usually come only after regulatory pressure or major press exposés, not as proactive ethical decisions.

So now it is up to the courts to rule on evolving social expectations of privacy and turn these arguments into legal standards anchored in decades-old statutes.

The Status and Outcomes of the Cases Against Meta as of Early 2026

By early 2026, several key outcomes are clear, even as other cases are still unfolding.

- The Cambridge Analytica user-profile litigation produced a $725 million settlement, now approved by the district court and defended on appeal against objectors challenging the amount and attorneys' fees.
- Facebook has paid a $5 billion civil penalty to the FTC and operates under a stringent, long-term privacy order that reshapes its internal governance and compliance obligations.

- Meta has reached significant settlements over facial recognition and BIPA claims and faces ongoing BIPA litigation in Illinois, where recent rulings confirm that Meta cannot use California choice-of-law clauses to sidestep Illinois' biometric-privacy regime.
- A $37.5 million settlement has resolved nationwide claims that Meta improperly tracked users' locations via IP addresses despite representations about location controls.
- The Meta Pixel healthcare litigation is moving forward, with courts rejecting Meta's broad dismissal arguments; now discovery and further motions will determine whether Meta faces trial or settlement in that arena.
- The FTC is actively seeking to tighten its order with Meta based on alleged noncompliance and to restrict Meta's ability to monetize minors' data, a significant looming threat to the company's future business model.

In response to this steady drumbeat of litigation and enforcement, Meta has incrementally changed its behavior by turning off or narrowing some features, implementing more compliance infrastructure, and publicly embracing the language of "privacy by design."

Yet Meta's fundamental economic engine remains the same: collect as much behavioral data as the law allows or appears to allow, transform it into predictive models, and sell targeted access to those predictions.

In turn, these cases illustrate this central battle between a company and the laws regulating it. When a company's profit model depends on invading privacy, lawsuits and fines become not just occasional costs but an ongoing operating expense. The law can slow Meta down, reshape some edges, and compensate some victims, but unless the underlying incentives change, privacy will always be something the platform gives back only under duress. And the same dynamic can affect its other legal and consumer battles, such as the claims that Meta's platforms are contributing to teen addiction. Meta will do what it can to maintain and maximize its profits, and it will do as little as possible to change its practices contributing to teen addiction where those changes threaten to reduce its bottom line.

In short, Meta faces a broader constellation of privacy cases and regulatory fines beyond the Pixel case. Together, all these cases show what Meta has already been forced to change and how much any further changes still depend on stronger law and enforcement in order to protect individual privacy and protect consumers from other ad and profit driven harms.

What Government Should Do to Protect Privacy

Given all of these cases against Meta for privacy violations, the government should establish some concrete policy steps to protect individuals from Meta-style surveillance. Current cases expose the gap, since U.S. privacy protections are fragmented by sector-specific laws (HIPAA, children's privacy, finance) and state experiments (GDPR-inspired comprehensive laws, BIPA-style biometrics), while general-purpose social-media data practices are only partially covered.

Some government actions that would directly address the patterns in Meta's litigation history and make changes to help consumers include enacting the following policies:

1. **A comprehensive federal privacy law with strong baseline rights**
    - Create clear, nationwide rights to access, correction, deletion, and data-portability for all personal data, not just specific sectors.
    - Require "data minimization" and purpose limitation by default, so that companies should only collect what is reasonably necessary for clearly disclosed purposes, with heightened standards for sensitive categories like health, biometrics, children, and an individual's precise location.
2. **Real consent standards and bans on dark patterns**
    - Define "consent" as specific, informed, and freely given, not buried in a long terms-of-service document.
    - Prohibit misleading interfaces or coercive designs that steer users toward oversharing, such as forcing a user to agree to cross-tracking in order to use basic features).
3. **Sensitive-data no-go zones and use-based limits**
    - Ban the use of health, biometric, and precise location data for targeted advertising, regardless of nominal consent, except under tightly regulated conditions.
    - Expand FTC and state authority to treat the misuse of sensitive data as an unfair practice even in the absence of a breach.
4. **Stronger enforcement and personal accountability**
    - Give the FTC clearer statutory authority and civil-penalty powers for first-time privacy violations, so that companies can forfeit profits from illegal acts and be required to make structural changes in their systems.
    - Extend the "personal certification" model from the $5 billion Facebook order requiring executives to certify privacy compliance or face individual liability to large platforms generally.
5. **Biometric and tracking-technology regulation nationwide**
    - Use the BIPA and Texas biometric settlements as templates for a federal biometric-privacy policy, requiring explicit, written, revocable consent and meaningful remedies, including statutory damages.
    - Regulate cross-site tracking tools like pixels and software development kits (SDKs) with clear rules about where they can be used (such as prohibiting them on patient or school portals) and what categories of data they may not capture.
6. **Stricter rules for minors and vulnerable groups**
    - Build on the FTC's effort to bar Meta from monetizing minors' data by codifying age-appropriate design and data-use rules for all large platforms, including establishing high-privacy settings and bans on behavioral advertising to children.
7. **Mandated transparency and researcher access**
    - Require large platforms to provide audited, public transparency reports and controlled data access for independent researchers, so that harmful data practices and dark-pattern designs are discovered earlier, not only when whistleblowers or lawsuits surface them.

What Individuals Can Do as Users and Citizens

Aside from seeking any redress from the government, individuals can take certain steps to raise the cost of abusing our data from Meta or other platforms.

- **Limit what you share.** When a platform makes it hard to find or change privacy settings, you should treat that as a red flag. Limit what you share, turn off ad personalization where possible, and avoid logging into third-party sites with Facebook.
- **Reduce the worst data flows.** Avoid connecting health, finance, and kids' activities to Meta products if you can. Use different browsers or profiles for sensitive tasks (such as using one browser for health and finance activities another for social media), which makes cross-site tracking less accurate.
- **Support institutions that say "no".** Ask your healthcare providers, schools, and employers if they run tracking pixels or social plugins on portals and internal sites. Reward those that remove them and pressure those that do not to make changes, which is what drove some hospitals to settle and change their web setups.
- **Back strong privacy laws and candidates.** Since privacy protections follow political pressure, support local and national candidates who talk concretely about comprehensive privacy law, biometric limits, and children's data protections, not just making vague "we care about privacy" statements.
- **Join or support civil-society groups.** Because organizations focused on digital rights can sue, lobby, and act as watchdogs in ways that individual users can't, supporting them amplifies your influence far more than making individual setting changes.

Regulating Meta and Platforms Like It

To sum up how to deal with privacy violations and other social media harms, any regulation should treat Meta as an infrastructure for attention and data. So any regulation should:

- Treat Meta and similar firms as "systemic" digital platforms, subject to enhanced obligations around privacy, security, and transparency, analogous to how we regulate important financial institutions.
- Tie continued operation in sensitive markets, such as youth services and health-tools, to audited compliance with high privacy standards.
- Make repeat major violations grounds for structural remedies, such as the forced divestiture of data-intensive business lines, bans on particular data uses, or court-supervised monitorships.

Such transformations in operations to protect personal privacy and stop other user harms like teen addictions are needed, rather than just bigger fines for violations, which can be treated as the cost of doing business. Rather these regulations should threaten the business itself unless its structure and systems are changed to protect consumers. This way, through changed ways of using social media, organized civic pressure, and serious lawmaking, we can change what the business model can do and limit who it harms by making it more accountable through both individual action and structural reform.

A chart summarizing the cases and their outcome or status is on the following page.

Case / action	Core issue	Main law(s) invoked	Outcome / status (early 2026)
FTC $5 billion order	Misleading privacy practices, violating 2012 order	FTC Act, 2012 consent order	$5 billion penalty; heavy governance/audits, ongoing oversight; FTC now seeking tighter limits, especially on youth data.
Cambridge Analytica / user-profile MDL	Sharing user and friends' data with apps, including political firms	Consumer-protection, contract, privacy torts	$725 million class settlement, final approval granted; Meta defending against objector appeals.
Location-tracking class action	Using IP-based location despite "control" promises	State privacy and consumer-protection laws	$37.5 million settlement; case closed.
Texas biometric (CUBI)	Face-recognition scans without proper consent	Texas CUBI, Deceptive Trade Practices Act	$1.4 billion settlement; ongoing project-notification obligations.
Illinois BIPA facial-recognition cases	Tag-suggestions facial templates without BIPA consent	BIPA (Illinois)	Large settlements and shutdown of tag suggestions; new BIPA cases (e.g., Hartman) continue against Meta.
Meta Pixel healthcare MDL (Meta as defendant)	Intercepting patient web and portal activity via Pixel	ECPA, state wiretap, contract, consumer-protection	Motions to dismiss denied; discovery and class-certification briefing ongoing.
Hospital-specific Pixel suits (e.g., Marin Health, North Bay)	Hospitals sharing PHI with Meta and others via Pixel	State privacy, consumer-protection, contract	Provider settlements (millions in cash plus commitments to reconfigure/remove tracking).
EU/Irish DPC GDPR enforcement	Legal basis for ads, password/data-security issues	GDPR	Multiple fines in the hundreds of millions of euros.

WHAT THE GOVERNMENT SHOULD DO

From Fines to Subpoenas: What Washington and the States Can Do

The federal government and state governments both have broad authority to force Facebook (Meta) to correct its entrenched problems with privacy abuses, account lockouts, scammers, bots, and mistreatment of advertisers, but they have not yet used that authority aggressively enough. Stronger enforcement, new legislation, and coordinated investigations could reshape how Facebook operates in the United States.

Why Government Action Is Needed

It's needed for the following reasons:

- Facebook has a long record of privacy violations, including the Cambridge Analytica scandal and repeated breaches of a 2012 FTC privacy order, leading to a record 5 billion dollar fine but ongoing disputes over stricter limits.
- Users continue to report being permanently locked out after "video selfie" or selfie verification reviews, often with no clear appeal path or human support, effectively erasing their accounts and access to contacts, groups, and business pages.
- Advertisers report ad budgets being drained by bots and fraudulent clicks, phishing scams triggered by running ads, and difficulty getting refunds when Meta counts clearly fake engagements as "legitimate."
- Facebook's growth has slowed or plateaued in mature markets while TikTok, Snapchat, and other platforms rise, shifting power but not reducing Facebook's structural power over communities, small businesses, and political discourse.
- These problems are outlined in detail in the book *What's Wrong with Facebook.*

What Congress Should Do

Congress should do the following:

- Pass a comprehensive federal privacy law that sets clear limits on how platforms collect, combine, and monetize user data, including bans on repurposing security data (like phone numbers or ID selfies) for advertising or profiling.
- Codify strong user rights: Face should be required to do the following: provide clear notice, data access, correction, portability, and deletion. It should provide fast human-reviewed appeal rights when accounts are disabled or monetized pages are taken down; and provide statutory damages when platforms ignore these rights.
- Require platforms over a certain size (including Facebook) to maintain independent, auditable systems for fraud detection in advertising, and to provide timely, itemized refunds for invalid traffic, with penalties for systematic ad fraud or failure to investigate complaints for problems with advertising.
- Clarify that "dark patterns," deceptive consent flows, and hidden tracking constitute unfair or deceptive practices under federal law, making it easier for the FTC and state AGs to sue over manipulative interface designs.

Role of Committees, Hearings, and Subpoenas

The House and Senate Committees should do the following:

- House Energy and Commerce and Senate Commerce Committees should hold recurring oversight hearings on Meta's compliance with privacy orders and its handling of scammers, bots, and account lockouts, rather than one-off headline hearings.
- Committees should issue subpoenas for internal documents on:
    - How "video selfie" and other automated verification tools are used, error rates by region, and appeal pathways.
    - Internal estimates of ad fraud, bot traffic, and refund rates, and any evidence that Meta knowingly profits from invalid traffic.
- The House Judiciary and Senate Judiciary Committees could examine whether Meta's market power and conduct toward advertisers and small businesses — such as refusing to refund obviously fraudulent engagement — warrants antitrust remedies or structural separation of business lines.
- Congress should mandate periodic public transparency reports from Meta, subject to perjury liability for executives, covering privacy incidents, account terminations, and ad fraud metrics.

FTC, DOJ, and Federal Enforcement Tools

- The FTC should treat repeat violations of its existing orders as grounds for escalating remedies, including higher civil penalties, mandated product changes, and bans on specific data practices (such as monetizing minors' data or using certain sensitive signals for ads).

- When Meta launches "kitchen sink" constitutional challenges to evade new privacy limits, the DOJ should robustly defend the FTC's authority and pursue parallel enforcement where criminal violations, obstruction, or false certifications are found.
- The FTC can require:
    - Independent third-party audits of ad fraud and bot traffic on Meta's platforms, with summaries made public.
    - Independent audits of account-verification systems, including selfie checks, to measure wrongful lockouts and bias across demographics.
- The DOJ and FTC can explore antitrust remedies if Meta's dominance allows it to offload the costs of fraud and privacy harms onto users and advertisers while competitors struggle to gain fair footing.

What State Governments Can Do

- State attorneys general can bring actions under state consumer protection laws against deceptive advertising metrics, mishandled refunds, and misleading claims about support and appeals when accounts or pages are disabled.
- States can pass privacy laws modeled on or stronger than California's laws, extending rights to transparency, correction, and deletion, and restricting how platforms handle biometric and ID data used in selfie or video verification.
- Legislatures can require platforms to provide:
    - A clear, prompt appeals process for account bans, including human review for users whose accounts are tied to livelihoods or essential community roles.
    - State-level reporting on scam prevalence, bot activity affecting local advertisers, and response times to user complaints.
- States can condition public agency advertising and official use of Facebook on minimum standards for fraud control, privacy compliance, and user redress, leveraging public contracts to push improvements.

Penalties, Restrictions, and Structural Remedies

Congress and state governments can provide for the following serious penalties, restrictions, and structural remedies:

- Monetary penalties: Congress can raise statutory caps and streamline penalty calculations so that large platforms face fines that scale with their global revenues for each day of noncompliance, making repeat violations economically intolerable rather than a cost of doing business.
- Operational restrictions: Lawmakers and regulators can prohibit Meta from deploying high-risk systems, such as selfie-only verification or fully automated ad review, without validated error-rate benchmarks, appeal mechanisms, and public documentation.
- Advertising restrictions: Regulators can require Meta to credit advertisers automatically when traffic is classified as invalid by independent auditors, and to pause campaigns when bot activity crosses defined thresholds, rather than continuing to bill advertisers.

- Data minimization and purpose limits: Statutes and orders should tightly limit what data Facebook can collect and how long it can retain it, especially for sensitive identifiers used in verification, creating strong incentives to design safer systems up front.
- Personal accountability: Senior executives could be subject to individual civil and, in serious cases, criminal liability for knowingly false compliance certifications or willful neglect of mandated privacy and security programs.

Summing Up

By using these tools together — investigations, subpoenas, fines, structural restrictions, and clear statutory rules — federal and state governments can push Facebook to fix deep-seated problems that it has repeatedly failed to address on its own, even as users and advertisers turn to competing platforms and Facebook's user base declines.

ARTICLES ON WHAT'S WRONG WITH FACEBOOK

Writer-Film Producer with Books and Films on Avoiding Scams Loses Scam Group to Facebook Scammer

I thought I knew how to be cautious and avoid scams after writing three books and producing two documentaries about scams and how to avoid them. But I ended up losing my Facebook account and two groups I created – "Scammed" with over 4800 members featuring different scams in the news, and "The AI Revolution and Writers and Artists" with over 500 members featuring the latest developments in AI affecting writers and artists. After I got back my email which Facebook removed from its system, I restarted the groups under new names: "Cons and Scams" and "The AI Revolution and Writing, Arts, and Business," over two dozen scammers joined these accounts and immediately offered help in recovering money or Facebook accounts, though I quickly blocked them and deleted their posts.

I was initially inspired to create these groups, since I wrote three books about scams: *The Big Con* about a book-to-film scam I discovered, *I Was Scammed*, about several dozen scams to avoid; and *Scam Story*, about a scammer who tried to scam me by claiming to be a PayPal rep. These experiences turned into two just released films by Gravitas Ventures: "Conned: A True Story" and "Con Artists Unveiled." Then I began working with AI about a year ago, and I created over two dozen adult and children's books using A, and I wrote and produced a film to be released next March: "The AI Revolution," also by Gravitas Ventures.

Here's how I got scammed on Facebook by the e-mail replacement scam; how to avoid it if you have a Facebook account; and the Facebook group you can join to learn about and discuss all kinds of scams, "Cons and Scams." The other group for those interested in AI is "The AI Revolution and Writing, Arts, and Business."

The Facebook email replacement scam began on Monday, October 16 when I got a notice posted on my Facebook account that an email was just added to my account. I thought it was just one of those warning emails from a scammer urging you to open it, so they can gain access to your computer or lead you to think they're going to help you so you give them information that leads to identity theft, access to your bank account, or getting money from you. The scenarios vary but the scams start with the scammers engaging you in an email exchange or phone call.

At first, after I ignored the notice, nothing seemed to happen. So I thought all was fine. Then, later that day and the next morning, I got three more warning notices, so finally I decided to open the notice but not click or respond, though this time the notice really did appear to come from Facebook, since it addressed me by name and included my city when it said, "Hi Gini. This is to let you know that the email address something@example.com was just added to your Facebook account," followed by the time of this change which came from my city and from Chrome. Then, there was a link to click to say: "This wasn't me," followed by the statement that "If this was you, you don't need to do anything. Thanks, The Facebook Security Team." The notice ended with Meta's name, logo, copyright notice, and the reference: "Attention: Community Support."

After I clicked "This wasn't me," I just had to prove I was me, which turned into a three-hour struggle, in which nothing seemed to work. One problem was that my account password no longer worked, since the scammer who had changed my email also changed the password. As a result, if I said "I forgot my password" and asked for a new one, that would go to the scammer's email, because as Facebook advised me: "We'll send you a code to your email. We can send a login code to: s*****@e*********.com and then you can "Log in with password". But unfortunately, that was the email of the scammer who took over my account.

Though I could try another way by sending a request for a new password to my phone, since I had two-factor authentication on the account, designed to make it more difficult for someone to hack my account, this was an old cell phone number because I set up the account over 10 years ago. But now I had a new phone, and I couldn't change the number.

I looked for a way to talk to someone at Facebook to explain the situation, but the chatbot only gave me canned answers on how to recover a Facebook account, change or reset your password, or learn what to do if you can't reset your password because you can't access the email address or phone number. When the chatbot said, "Was this helpful?" and I said "No," I only got more canned answers about recovering my account. And when I looked up a number for Facebook, I only got a voice mail referring me back to the chatbot, because, as I later learned, Facebook offers no personal support. It's all "ask the bot" or nothing.

Finally, after about two hours of this, I clicked a "Try another way" button that gave me a final option to prove who I was. I had to hold up my driver's license, my passport, or other acceptable government document with my picture and name in front of my webcam. So over the next hour I tried holding up my driver's license and then my passport. As I did, an orange-colored square or rectangle occasionally appeared, as if it was searching to find my license or passport. From time to time I got the message, "hold your document closer to your webcam," which I did, and then I got instructions to hold it so all four corners showed and to be sure there was enough light. Repeatedly, that's what I did, and from time to time I moved my fingers into different positions or adjusted the lights behind me to shine more light on the webcam. But each time that the camera behind the orange-colored rectangle seemingly snapped my driver's license or passport, I repeatedly got the message, that this was unacceptable, so "retake your photo." So again and again I did so, holding up my driver's license or passport, so the camera could retake a picture again and again, though sometimes that meant holding up the card for a couple of minutes, while the webcam went through its paces of finding the card, only to tell me to retake the photo.

Finally, I gave up, figuring I needed help from someone who could hold up a flashlight for me, so I could shine my light on license or passport if that was the problem, since I couldn't both hold it up and shine the light.

However, about two hours later, when I thought I could rig up a lamp to try again, it was too late. Now when I got the message to "Find Your Account. Please enter your email or mobile number to search for your account" I got the notice: "You're temporarily blocked: It looks like you were misusing this feature by going too fast. You've been temporarily blocked from using it. If you think this doesn't go against our Community Standards let us know." But how could I do this, since I couldn't prove I was me?

Needless to say, I felt incredibly frustrated. Though a friend who knew computers offered to help me the following day, now it was too late, since a few hours later, I got a notice in my email for my Facebook account that announced, "Your account has now been removed."

So the scammer had successfully taken over my account, though it wasn't clear why, since over the next week, the scan and AI groups I had managed continued to exist, though there were no new posts from the scammer and people could still join, although no one could post anything.

Meanwhile, since my original email was removed by Facebook, I discovered that I could create a new account under this email and decided to restart my groups under a new name. So "Scammed" became "Cons and Scams," while "The AI Revolution and Writers and Artists" became "The AI Revolution and Writing, Art, and Business." And now they are already starting to get new members, while I know that no one has posted anything in the old accounts, because I'm still a member and can see the number of members in the account. I was even able to rejoin under my original email, though when I tried to post an announcement that I was no longer running the group and that I set up a new group, my post remained pending.

Ironically, it's probably just as well that the old "Scammed" group died, because after the group hit over 500 members, almost all the new members were scammers offering services by themselves or others to recover money or accounts for people who were scammed. So I repeatedly deleted their posts and blocked a few scammers who tried reposting their messages again and again.

Another irony is that after I set up the new AI Revolution group, about two dozen of the new members who joined tried to post "recover your money or account" scams, though I quickly reported the scammers to Facebook, blocked them from the site, and deleted their posts. So how did they find me so quickly? I guessed that the posters must have been the original scammer using different emails, since I had submitted a post to announce my new site. But since the post was pending, no one could see it except the scammer who took over the site.

In any event, this experience has been an eye-opener about another dangerous scam that I never saw reported anywhere – the scam that hackers can take over Facebook groups by changing the email of the group. Then, if you aren't able to respond quickly enough to prove it wasn't you, they can take over your site and perhaps use the site or seek ransom if someone wants to get it back. Or maybe they can use the information about you or others gained on the site to do something else to cause harm.

To protect yourself, make sure that your cell phone to verify yourself is up to date, and if asked, quickly provide the necessary documentation to prove you are you.

Facebook's Face-Off: How Facebook's New "Video Selfie" Is Locking Out Real People—and Raising New Risks

Facebook is now using "video selfie" identity checks to prove that an account belongs to a real person, but these checks are locking out real people, including me, after I had an account for over 20 years.

The Vast Number of People Affected

Precise totals are hard to pin down, but millions of people are affected. A U.S. local tech segment reported that "some 110 million people" get locked out of Facebook accounts annually (not all due to selfies), a number that suggests account-recovery frictions affect vast swaths of users. Meanwhile, Meta has said it deleted roughly 10 million accounts in the first half of 2025 for impersonation, spam, and fake engagement—evidence of an aggressive enforcement environment that likely increases false positives.

How It Works – Or Doesn't

To provide your identity, you are asked to record a short selfie in a circle while you turn your head up, right, or left according to a series of random arrows, Additionally, Facebook may also match the face to photos on your profile if you have your picture there, though some people don't have any picture or may have a logo or other image. Then, Facebook uploads a file with your image, sees if it meets its criteria for being an acceptable image, and if not, you may be asked to retake it or receive a decline.

Meta has set up this system to avoid people setting up fake accounts, but many real people may find they can't pass the test, and even people with longtime accounts may suddenly find they have to prove they are a real person because they have used their account in a way that draws Facebook's attention. Then, they have to go through this selfie process, and if they can't pass the test, too bad. They are suddenly without their account, and they may find it difficult or impossible to appeal the decision, and there doesn't appear to be anyone you can speak to personally.

My Experience with Facebook

That was my experience – and that of many others.

In my case, the loss of my Facebook connection began when I signed up for a private Facebook group and Facebook wanted to send a verification code to my email. But after I invited Facebook to do so, I discovered someone else's email on my account, indicating it had been hacked, but I never realized it before, since I had used my account regularly without signing out. So now I couldn't sign back in and lost that account.

Yet I had another account under my company's name, with my company's initials for the logo. I hadn't used the account for five years, but at least I could sign-in and use it for connecting to several Facebook groups I belonged to. So I began adding personal contacts, and initially Facebook helpfully suggested people to invite, though I could only invite up to four or five people a day, or I would get a notice that "You can't use this feature now." At least the invite feature seemed to work well, as I built my account up to 80 participants by inviting people who I knew or had two or more other friends on Facebook.

But then I got the dreaded "take a selfie" request, because I clicked on one of the people that Facebook suggested I invite as a friend, and I was advised that "It doesn't seem like you know this person." Then, I made another two invites. But on the third one, I got the "You can't use this feature now" response, and moments later, I got a notice that 'Your account has been suspended for inappropriate activity." Now if I wanted to appeal, I had to take a selfie. And that, I found, proved to be impossible. Repeatedly, I tried to take a picture by positioning and repositioning myself, moving a little further away from my computer each time, so my face would fill the circle and I could better turn right and left for a side view. But each time, I got the "retake" response, and I was left wondering if I turned far enough or fast enough to be in the proper position.

I finally gave up trying to take a picture and restore my account myself. Initially, I had a VA create an account for me with another one of my emails, and I used a second computer to access it, since the original one was linked to my declined Facebook account. The new account worked for a time. However, the VA was in India, so I had a news feed with news and images from India, though I was able to post a few messages to say, "Hello, I'm here," Unfortunately, this other computer had a sporadic Internet connection, because it was a Win10 computer, and at times, I couldn't connect to my Facebook account and had to sign in again. For a few days I could sign in without the selfie requirement, perhaps since the account was set up in India where a personal verification was less needed. But after a series of internet outages and failed attempts to sign in, I got the dreaded selfie prompt. So once again, I couldn't access my Facebook account.

After that, when a friend tried to set up an account using my phone number and a different email, that didn't work either. After I briefly put in a code number on my phone, Facebook still wanted a selfie, and after my friend's girlfriend tried taking the pictures and at first passed the test, the account was declined, though, as is commonly the case, Facebook gave no reason for its decision to ban an account.

I even tried writing to appeal, sending my appeal to several emails listed on Google for appeals: appeals@fb.com, support@fb.com, and disable@fb.com, explaining what happened.

But that effort to appeal didn't work either, since the support@fb.com account was full, so the message never got to Facebook, and I didn't get a response from the other emails.

Thus, that decline coupled with Facebook's lack of response was the final straw, which led me to turn to other social media accounts – Instagram, X, TikTok, and LinkedIn, and to investigate whether what I experienced with Facebook was experienced by others.

The Many Problems of Others with Facebook

As I quickly learned, there were many group chats on Reddit and other platforms where people complained about losing their accounts due to the new system. Commonly they lost their accounts after they were asked to verify themselves with a selfie and couldn't do it or didn't want to, because they felt taking the picture was an invasion of their privacy, believed the images could be used for devious purposes, or otherwise felt so angry about the system that they decided to leave Facebook.

Here is a sample of their responses.

"Facebook is now forcing me to provide a video selfie to get back into my account. Something seems fishy

"FACBOOK has locked me out of my account until I provide an official facial recognition video selfie…is anyone dealing with this? I just feel frustrated.

I created an account and it IMMEDIATELY got suspended and now Facebook is asking me to take a selfie of myself?!?!?"

"Facebook wants me to upload selfie to create an account. Is this normal now?"

"Facebook requires me to upload a video selfie to prove I am human, but I am on a pc and do not have a camera. how do I bypass this?"

"My 20-year-old Facebook account was suspended, and Meta was demanding "video selfie"

"Facebook is now requiring me to take a video to open new account. What the heck are you doing. Facebook?!"

"Why is Facebook asking my 78-year-old mother who doesn't have a webcam for a video selfie?"

"What in god's name is this? I decided to make a Facebook account for the first time, and they asked me make video identity tests and a lot of other crap."

"My new Facebook account asked for a video selfie after 30 minutes. What should I do?"
"New account video selfie verification not working? Anyone else having this issue?'
"Facebook disabled my account, asked for a verification selfie, then permanently deleted my account."
"I failed the video selfie verification; will I ever get a second chance?"
"Facebook declined my verification selfie video and banned my account."
"I was asked to upload a video selfie and then what? My account is still suspended and nothing is happening."
"I was suspended over the lack of a full facial selfie in my profile picture? That is the only thing I can think of at this point."
"Meta thought I was younger than 13. But I'm 37 and all my accounts were suspended."
"My Facebook account was suspended with no way to appeal?
"If your Facebook account got suspended recently, you're not alone let's do something about it."
"My Facebook account got suspended. Please help because I don't know how to proceed from here."
"I opened a FB account, and after 2 days it was officially suspended, what gives?
"Has anyone questioned why they are asking us all for a video selfie?
"I was told that your account was suspended because your linked Meta account doesn't follow our rules."
"Facebook suspended and will disable my account in 180 days if I do not do anything. My response to Facebook and their new rule for a fucking damn selfie. "

Two Detailed Responses to the Facebook Selfie

After looking at all these short complaints, I saw two responses that seemed very much like what I wrote about what happened to me.

As one man stated:

"Looks like a lot of people are in the same boat. I've been using FB for around 20 years, with no warnings or anything before. I don't post much on my personal account, nothing inflammatory, and I'm the admin of a number of large groups, some of which have 50k+ people. The only "reason" I can think of for my ban is that I often post anti-AI stuff on FB and report a lot of offensive posts/comments as well as AI adverts. Anyway, I received the following email:

"Your Facebook account has been suspended. This is because your account, or activity on it, doesn't follow our Community Standards on account integrity. If you think that we've suspended your account by mistake, you'll have 180 days to appeal our decision. If you miss this deadline, your account will be permanently disabled."

Another man complained:

"I got this notice from Facebook, and I don't understand why after having an account for 20 years. Here's what it said: "We disabled your account…We reviewed your account and found that it still doesn't follow our Community Standards on account integrity and authentic identity…You cannot request another view of this discussion…To learn more about the reasons we disable accounts visit the Community Standards…You may have the right to challenge our decision in your country's courts, or refer our decision to a certified dispute settlement body."

The Major Reasons for Common Complaints

These are just a sampling of the many complaints about Facebook's new selfie video verification process which I found online. As numerous current and former Facebook users reported, the process is frequently flawed, often rejects facial uploads without clear reasons, and makes it virtually impossible to appeal or get support. Many users report that their accounts were locked, suspended, or disabled after failed video verification attempts, and they were frustrated by the lack of recourse, with some believing the process is a way to force users into paid subscriptions. The issues are particularly prevalent for users with older accounts, new account registrations, and those trying to advertise. And individuals who lost their accounts commonly complained about the difficulty of contacting support without an active account. Here's a summary of these common complaints:

- **High rejection rate:** Many users report that their video selfies are rejected multiple times, even when they are clear and follow instructions.
- **Lack of appeal:** There is a strong sentiment that rejected verification attempts are final, with no option to appeal the decision.
- **No support:** It's difficult to get support to resolve issues, as many users are locked out of their accounts and cannot access the help center.
- **Account suspension/lockout:** Users find their accounts suddenly locked or suspended, with the video selfie being the only way to regain access, leading to frustration and potential loss of data.
- **Spam/scam concerns:** Some users are worried about Facebook collecting facial recognition data through the process, with fears it could be used for other purposes without consent.
- **Forced subscription:** Some speculate the process is designed to pressure users into purchasing a paid account to get support or unlock their accounts.

- **Technical glitches:** The upload button doesn't work, and videos won't upload for many users.
- **New account quickly disabled:** As a last resort, some users have created new accounts, but even these are sometimes disabled immediately after creation.

Why the Selfie Doesn't Work for Many Legitimate Users

There are three main reasons the selfie approach doesn't work well for many users:

1. **A problem with the system:** It is difficult to take a selfie for various reasons, such as poor lighting, camera quality, or minor appearance changes that prevent matching.
2. **A lack of explanation from Facebook about what is wrong:** People rarely get a clear reason for rejection, and they don't have a way to speak to a human to resolve the problem, making lockouts feel arbitrary and permanent.
3. **No alternative way to get back into the system:** For many locked-out users, they can't get back in, since users have no other way to prove their identity.

Why Facebook Is Seeking These Profiles

Meta's public explanation for creating this new facial recognition requirement is that the company is testing more automated recovery options to fight scams and speed up access restoration for locked or compromised accounts. In an October 2024 update, Meta said its video-selfie verification would roll out as an optional method in more regions for people trying to regain access. Their stated goals are to combat impersonation, reduce fake IDs, and make recovery "faster and easier."

Meta is doing this because it faces intense pressure to curb scams, deepfakes, and large-scale impersonation campaigns. It claims selfie verification is a way to "combat scams" and "help restore compromised accounts," and its help center says video selfies help "prevent fake IDs." For public figures, Meta has also tested face recognition to spot impersonation in ads. The push reflects broader platform and regulatory trends: more automated checks to prove you're a real human.

But in practice, a wave of users reports the system is glitchy—and once it fails, appeals can stall for weeks with no human follow-up or there is a lack of response. On Reddit, multiple long threads describe accounts being locked until a "facial recognition video selfie" is provided, only for the verification to be rejected and the account to remain disabled. In several discussions, longtime users say they never recovered access after completing the video step, fueling perceptions that automated moderation—not a human—made the final call.

The Failure of the System to Get Rid of the Bad Actors

The problem with Facebook's stated goals for the system is that bad actors are already trying to exploit or bypass video-selfie checks. Underground tools openly advertise "video stream spoofing" to pass Facebook's selfie verification for multiple accounts—a reminder that any scam-stopping or other virtual system can quickly become a cat-and-mouse game the moment it's deployed. Forum posts and Facebook group discussions also circulate tips on working around the selfie prompt, underscoring that determined abusers quickly adapt.

In short, Facebook's selfie verification attempts to solve a real problem—mass impersonation and account theft—but it suffers from ineffective automation, limited recourse, and growing evidence of spoofing tools that erode its deterrent value. Until Meta offers reliable human review and transparent error remediation, the system will keep locking out real people while motivated scammers probe the gaps

How You Might Be Able to Recover Your Account

Although these approaches don't always work, here are some suggestions I received from various sources:

1. **Start with the official portals—on a known device.** Use facebook.com/hacked from a phone/computer you've used before; logins from "familiar" devices can help. If you see the selfie prompt, follow the motion cues precisely (neutral background, bright front-facing light, clear camera image).
2. **Look for alternate routes (if shown).** Some areas may allow government ID upload instead of a video selfie; if offered, it can be more reliable, though not all users see this option.)
3. **If you're a paying subscriber, try Meta Verified support.** On Facebook/Instagram, subscribers may get access to chat-based help from within the app. Results vary, but it's one of the few paths toward a human.
4. **Don't use "unlock" services.** Third-party "recovery" services may send you an offer, but they often harvest your data or don't work.
5. **Harden your broader footprint.** Turn on 2FA for critical accounts, store backup codes safely, and avoid mass friending bursts right after a recovery, since these behaviors can retrigger risk checks. Industry reports show scammers are constantly probing verification systems; minimizing risky patterns helps you avoid red flags that trigger a self-prompt.

Summing Up

Facebook's selfie verification system aims to curb fakes and speed restoration—but the current system is opaque, error-prone, and difficult to appeal. For many, including this author, the lack of a reliable human escalation path turns an identity check into a permanent lockout—and a reason to leave the platform. Until Meta pairs automation with transparent recourse and consistent human review, video selfies will keep punishing real people while determined scammers have been able to adapt to stay on Facebook.

Facebook's Decline and Its "Video Selfie" Problems

Facebook is in a state of decline, characterized by a loss of daily users and ad sales, and contributing to that decline is its "video selfie" identity program. Not only is Facebook suffering from decreased engagement, a negative perception among a younger audience, and a drop in ad revenue, but requiring a video selfie to prove one is a real person has caused a growing number of users to drop out. That's because the system is locking out real people since the picture taking feature doesn't work very well. That even happened to me after I had an account for over 20 years, and my multiple attempts to get the account back didn't work. This article highlights the main points about Facebook made in the previous article.

The Loss of Facebook Users

This loss of users is huge. While Facebook has about 2.11 billion daily active users and around 3.07 billion active users each month, according to internet stats, it has recently lost about 500,000 daily users according to the *Washington Post*. The result is declining stock prices, lower ad revenues, and a growing number of users expressing dissatisfaction with Facebook's policies and practices.

The Vast Number of People Affected

As previously noted in my last article, Facebook's selfie-program is a major source of this loss of users affecting millions of people. A U.S. local tech segment reported that "some 110 million people" get locked out of Facebook accounts annually, many due to selfies. According to Meta, it deleted about 10 million accounts in the first half of 2025 for impersonation, spam, and fake engagement — an aggressive enforcement environment that has made false positives in its selfie program more likely.

How the Self-Program Works – Or Doesn't

As previously noted, to provide your identity, you are asked to record a short selfie in a circle while you turn your head up, right, or left according to a series of random arrows. Then, Facebook uploads a file with your image, sees if it meets its criteria for an acceptable image, and if not, you may be asked to retake it or receive a decline.

Meta has set up this system to avoid people setting up fake accounts, but many real people can't pass the test. Even people with long-time accounts may suddenly find they have to prove they are a real person due to using their account in a way that draws Facebook's attention. Then, they have to go through this selfie process, and if they can't pass the test, they lose their account, may find it difficult or impossible to appeal the decision, and can't speak to anyone personally to restore their account.

How the Selfie Program Began and Led to Real User Lockouts

Facebook began testing the video selfie verification process to curb scams, deepfakes, and large-scale impersonation campaigns and fake IDs. But in practice, the system is glitchy—and once it fails, appeals can stall for weeks with no human follow-up or there is a lack of response. While it was used to restore access to accounts that were compromised or locked due to suspicious activity and to prevent bot creation, the program ended up locking out real users and raised other concerns. These main causes of user loss and frustration are these, according to reports on Google:

- **Failed verification leads to permanent lockout:** A high number of users have reported that after submitting a video selfie, the system either rejects it or fails to process it correctly, resulting in their account being permanently disabled. For many, this process has no clear pathway to appeal.
- **Technical glitches and poor support:** Numerous complaints cite technical issues where the video upload button fails to work or the process gets stuck in a loop, further frustrating users. In many cases, users find themselves with no access to human support to resolve their problem.
- **Loss of personal content:** When accounts are locked or disabled, users lose access to their personal photos, messages, and memories. This permanent loss has caused some to give up on the platform entirely.
- **Lack of transparency:** Many users are frustrated by Facebook's lack of clear communication on why their video selfie was rejected or why their account was disabled.
- **Privacy and data collection concerns:** The video selfie feature has fueled significant privacy concerns and distrust in how Facebook handles user data. Many users required to

submit a video selfie to access their account, fear that Facebook is collecting their biometric data for other purposes.

Other Factors Contributing to Facebook's Decline

While the video selfie program has led to Facebook's decline, some other factors are these:

- **Competition from other platforms:** Facebook has faced strong competition from newer, more engaging social platforms like TikTok and Instagram, also owned by Meta.
- **Changing user demographics:** Young people have been leaving Facebook for other platforms, changing the site's overall demographic.
- **Negative public perception:** Widespread media coverage of privacy violations, misinformation, and other controversies has damaged public trust in Facebook and led to a decline in user growth.

The selfie program adds to these other factors, helping individuals decide to leave Facebook.

Summing Up: Why the Selfie-Program Is Resulting in Declining Facebook Users

To sum up, Facebook's selfie program is contributing to its decline in users because the process is frequently flawed, often rejects facial uploads without clear reasons, and makes it virtually impossible to appeal or get support. Many users report that their accounts were locked, suspended, or disabled after failed video verification attempts, and they were frustrated by the lack of recourse. The issues are particularly prevalent for users with older accounts and new account registrations. And individuals who lost their accounts commonly found it difficult to contact support without an active account. Here's a summary of these common complaints:

- **High rejection rate:** Video selfies are rejected multiple times, even when individuals follow instructions.
- **Lack of appeal:** Rejected verification attempts are commonly final, with no option to appeal the decision.
- **No support:** It's difficult to get support to resolve issues, as many users are locked out of their accounts and cannot access the help center.
- **Account lockouts:** Users' accounts are suddenly locked or suspended, and the video selfie is the only way to regain access, but that often doesn't work, leading former Facebook users to feel frustrated and lose their personal data in their account.
- **Technical glitches:** The upload button doesn't work, and videos won't upload for many users.
- **New account quickly disabled:** As a last resort, some users have created new accounts, but even these are sometimes disabled immediately after creation.
- **Spam/scam concerns:** Some users are worried about Facebook collecting facial recognition data through the process, with fears it could be used for other purposes without consent.

The Failure of the System to Get Rid of the Bad Actors

Thus, while the system repeatedly fails regular users whose accounts are suspended or permanently disabled, bad actors have been able to exploit or bypass video-selfie checks. Underground tools openly advertise “video stream spoofing” to pass Facebook’s selfie verification for multiple accounts, and forum posts and Facebook group discussions also circulate tips on working around the selfie prompt. So determined abusers have been able to adapt and get through the system, while legitimate users are repeatedly shut out.

In short, Facebook’s selfie verification attempts to solve a real problem—mass impersonation and account theft—but it suffers from ineffective automation, limited recourse, and spoofing tools that erode its deterrent value. Until Meta offers reliable human review and transparent error remediation, the system will keep locking out real people while motivated scammers probe the gaps

How You Might Be Able to Recover Your Account

Although these approaches don’t always work, here are some suggestions I received from various sources:

6. **Start with the official portals—on a known device.** Use facebook.com/hacked from a phone/computer you’ve used before; logins from “familiar” devices can help. If you see the selfie prompt, follow the motion cues precisely (neutral background, bright front-facing light, clear camera image).
7. **Don’t use “unlock” services.** Third-party “recovery” services may send you an offer, but they often harvest your data or don’t work.
8. **Avoid some of the actions that might trigger a selfie request.** Turn on 2FA for critical accounts, store backup codes safely, and avoid mass friending bursts right after a recovery, since these behaviors can retrigger risk checks. Industry reports show scammers are constantly probing verification systems; minimizing risky patterns helps you avoid red flags that trigger a selfie-prompt.

Facebook's Declining Audience, Losing Ad Money, and Its Identity Program That Doesn't Work

I recently wrote about how Facebook was losing users, and one factor was its "take a selfie" program, supposedly to get rid of bots and fake users. And now there's another problem – advertisers who place ads, have their ads declined, and then they don't get back their money.

In the case of the self-program, the problem is that the program is getting rid of thousands of legitimate users. In the program, you take a selfie by positioning your head in a circle on the screen and moving your head up, down, right, and left in response to randomized arrows to prove you are a real person. Even though I tried do so over a dozen times, each time I was required to retake the video and finally gave up. Later, I learned that the program doesn't work very well, resulting in many users like myself being unable to prove themselves to the program or appeal to a real human to explain who they are.

Plus I just learned that many advertisers pay money for ads, but then the ad isn't approved and they don't get back their money. As one Facebook user, Phil Green, in the health care field told me: "Facebook allowed me to deposit $1,000 into my advertising account, but then disapproved my business website—effectively blocking me from using the funds I had already committed. Despite multiple attempts to resolve the issue, there's no accessible customer service channel to speak with a real person or get meaningful support. The platform's automated systems offer no explanation or path forward, leaving my account funded but unusable, and my business stalled without recourse."

In my own case, after I lost my personal account by not being able to provide an acceptable selfie, I reactivated a business account I hadn't used in 5 years, began posting on it, and invited people I knew or suggested to me by Facebook. But one day, after I invited someone Facebook recommended who had two friends I knew, I got a message that I couldn't use this feature now. A few minutes later, I got a notice that my account was suspended for improper activities against community standards.

Say what? I was only inviting people that Facebook suggested. Then, that experience led me to do some research about Facebook, its selfie program, and its practice of deleting accounts of individuals who found it practically impossible to appeal the process.

So all those problems, I realized, contributed to Facebook's declining audience, along with many former users, especially younger users, leaving it for other social media platforms.

Ironically, after deleting my accounts, Facebook sent me a notice about how it was personalizing accounts and ads. As Facebook explained, personalizing your experiences includes suggesting content like posts you may find interesting and reels to watch, along with showing ads more relevant to you. As Facebook wrote:

Updating how Meta personalizes experiences

Hi Gini Graham

On December 16, 2025, we're making changes to our Privacy Policy. Here are some details.

Personalizing your experiences

We'll start using your interactions with AIs to personalize your experiences and ads.

What this means

Personalizing your experiences includes suggesting content like posts that you may find interesting and reels to watch. It also includes showing ads that are more relevant to you.

Thanks,

Meta Privacy

So why was Facebook sending me this invitation, after they had prevented me from using my personal account due to a glitch-filled selfie-program so I couldn't prove I was me and then disabled my business account after I invited a person Facebook suggested to join my account? Didn't Facebook know they had disabled my account? And shouldn't they know that before inviting me to advertise and telling me about their latest privacy policy to protect me? Feeling frustrated by the whole experience, I sent off a letter to Facebook support, thinking it might lead someone to contact me about the confusion. As I wrote:

> "Why are you sending me this? You cancelled my account for no reason, presumably because I invited some people you recommended that I invite to become members of my account, and these were people where I already had one or two friends. You also invited me to take a selfie test, and though my face was within the circle, you kept asking me to retake it, and I finally gave up. It is ridiculous that you are sending me a notice about personalizing my account and my ads, along with a privacy notice about protecting my account, when you already took it away!"

Needless to say, I didn't get any response to my email, and my account remains closed. But since it has, this experience helped me think about all the reasons I didn't need Facebook and learn why Facebook has become irrelevant for many people, contributing to its decline. I realized that the main reason I continued to use Facebook was because three writing and film groups I belonged to had Facebook groups for members, though one was already thinking of moving its group elsewhere due to the problems I and others encountered in using Facebook.

Another reason I didn't need Facebook, I realized, is beauset I had created a group on Facebook called "Scammed," since I had written several books and produced two films on the subject. But while I sporadically posted links to articles on scams, the group's membership, now about 7300 members, was largely made up of scammers seeking people who had already been scammed for a "recover your funds"| scam. And Facebook, despite its efforts to clean up the site by disabling accounts, hadn't done anything to get rid of these scammers.

Finally, another reason I didn't miss leaving Facebook is that even though I had built up my personal account to 80 friends when it was shut down, people had routinely accepted my invitation to join, but after that, t I didn't have any further communication with them. So why did I need them as friends in my account? I frequently saw them at local events and could email or call them if I wanted to communicate with them.

In short, the more I learned about Facebook, the more I recognized different reasons why it was declining in popularity. It was becoming less and less relevant for many people, so they were dropping out at the same time that Facebook was deleting its legitimate users with its selfie program that didn't work very well and decision to suddenly disable accounts. And even legitimate advertisers were declined but then didn't get back their money.

In fact, Facebook's action to disable a swath of accounts that could affect legitimate users seemed like the current immigration crackdown that has been eliminating many immigrants who are documented or otherwise legitimately in the U.S. And the irony is that even after deleting these legitimate user accounts, Facebook is reaching out for more advertising by contacting people like myself whose accounts have been disabled. And perhaps Facebook is now making this effort to seek more advertising because the more its users decline, the more there is a decline in its advertising, too. Then, too, it would appear that Facebook is having problems with one department talking to another, in that those sending out information about personalization, privacy, and advertising aren't aware that they are sending this information to those who no longer have accounts.

But on the bright side, leaving Facebook has led me to become even more active in the social media spaces where I already have accounts or plan to set up new accounts on especially popular sites like Instagram, TikTok, LinkedIn, and X.

The Problem with Facebook Ads – From Scams to Losing Money

Facebook has long claimed its ad platform is an ideal engine for targeting the right users, setting up a budget with proven results, and delivering measurable returns. But for many small businesses and advertisers, reality appears quite different: campaigns get shut down, funds are not returned when ads don't run, clicks are driven by bots, and scam ads run along with legitimate ones. There are hundreds of complaints on Reddit and other sites about problems that advertisers experienced, and I began to research the problems with ads, after one associate, Phil Green, a health practitioner, told me:

"Facebook allowed me to deposit $1,000 into my advertising account but then disapproved my business website—effectively blocking me from using the funds I had already committed. Despite multiple attempts to resolve the issue, there's no accessible customer service channel to speak with a real person or get meaningful support. The platform's automated systems offer no explanation or path forward, leaving my account funded but unusable, and my business stalled without recourse."

Then I began discovering many other problems with Facebook advertising, as described in this first of a series of articles on these problems and what to do about it.

What Facebook Earns from Its Ads – Even Fraudulent ones

Facebook's earnings from advertising are huge. According to researchers, Facebook's parent company, Meta, generates its earnings primarily from advertising, with annual ad revenues reaching well over $100 billion. In 2024, Meta's global ad revenue was over $160 billion, a slight increase from the previous year, making up the vast majority of the company's total income. It draws most of its income from advertising to one of the world's largest social media platforms, with over 3.07 billion monthly active users.

Typically, Facebook ads cost from 50 cents to $3.50 per click, while the average CPC, cost-per-click, is 51 cents and the CPM, cost-per-thousand views, is $8.77, whether or not someone clicks on your ad. The cost of these ads varies depending on your industry, target audience, goal, and bidding strategy, with an average weekly cost of $35 to $125 per week. The ads work through an auction system where you set the highest amount you are willing to pay for clicks, impressions, or conversions, and you compete with other ads, in which the advertiser with the highest bid and most relevant ad wins the auction. Mostly commonly, advertisers use budget-based bidding in which they set a maximum budget for their ad campaign, and Facebook automatically bids up to that amount. You can manually set the amount you are willing to pay for a click, impression, or conversion, or Facebook can do it based on your budget and goals.

So that's the basic system, although it is rife with scammers, and there can be problems with Facebook ads going over budget or individuals losing money on ads that do not run. There are all kinds of red flags and reports of problems from advertisers who have lost money – sometimes hundreds or thousands of dollars.

Examples of Complaints

Here are some representative complaints from advertisers who have encountered a range of scams and financial problems:

Overcharges and Unapproved Billing

> "I cancelled an ad and three months later Facebook restarted it without telling me and hit my card for $600. I called in super pissed off and they eventually agreed to refund my money. I cancelled the account so they couldn't do that again."

> "As the year went on, these extra charges (from Meta) accumulated until now, where we're seeing thousands of dollars in unknown Meta ads charges per day…To be crystal clear, these are completely unrelated by my business' legitimate Facebook advertising. The transactions do not match anything in my Meta account, and I don't own any other businesses, so they are 100% fraudulent…I've been going back and forth with Meta's extremely unhelpful support. I provided a spreadsheet of every last charge which does not match anything in our ad account. They said that 'There is no suspicious activity in your ad account,' therefore they will not be refunding anything."

> "I set the daily budget for one set of ads at $5 a day and… another set at $1 a day. As soon as the ads began running, I started getting bombarded with $2 charges from Meta…and on the third day, the total was $24, so I turned the ads off…But I kept getting

charged… One of the problems is…the excessive charges caused my credit card company to flag the charges and now I can't use this credit card on the Meta ad accounts, so I had to switch payment methods…The bigger problem is I can't figure out why I'm being charged per day more than what I set the daily budget at. It should be $18 a day at most…I don't understand why I'm being charged more than I set the budget at."

Unauthorized Transactions

"I am writing to bring to your attention a series of unauthorized transactions that have been made from my bank account, labeled as payments from Facebook ads. These transactions, totaling over $20, occurred within a very short time frame, some even within seconds of each other.

"I previously owned a Facebook page, which has been inactive for several years. The page was deleted, and I have not accessed or posted anything on it since its deletion. Therefore, these charges for advertising do not make sense, as I have not authorized any such activity…

"It appears that my card information has been used fraudulently to facilitate these unauthorized transactions. To prevent further theft, I had to urgently block my card, which incurred an additional cost to issue a new one.

"I wrote tickets to support, and no one is responding…How can Facebook go so low to steal money."

False Charges

"I'm being billed by FB. Why? Looks like it's for ads but I don't do any advertising on FB. Just a personal account. The payment method is so I can send cash to my son.

"I went into Payment settings and stumbled into Billing & Payment. There I noticed some transactions in September of 2024. This looks like payments deducted from my payment method for ads. All of them in September. $18 dollars over 7 payments. What gives?"

"I rang my bank and stopped Facebook taking money altogether off me. These ads were back in August; they are scamming people. My ads were stopped months ago, no reason for the charges now when I already paid."

"Facebook straight up just stole money from me…As a VERY small business working from home… On Tuesday things were going normally…seeing it go up by about £5 or £10 as usual, then I …watched it jump up by about £60 right in front of my eyes. I contacted support right away and they admitted it was a glitch in their system and I should await updates.

"A couple of days go by and... they took their payment even though it was well over the amount I actually owed...I contacted them again and got the usual copy/paste response where they tell you someone will be in touch within 24 hours. Which brings us to today, I just got an email saying 'Unfortunately this case has been marked as closed. We would request you to open a new chat/email session and ask for a call-back, if the issue persists.'

"So basically, they've stolen my money and are ignoring me despite admitting they had a glitch that was causing this issue. I'm so angry with Facebook. Every time I've ever contacted their 'support' they've closed the case without ever resolving any issue I've had with them but this time they've gone too far."

Disabled Account and Continued Billing

"Yesterday morning, December 2, 2023, I wake up, and when I went to log in, it said that my account was disabled due to violating community standards and account of integrity of authentic identity. I cannot request another review of this decision. Then this morning, December 3, I get notifications that Facebook charged my account for $249.99, $250, and $250. For my ad's account that I CAN'T EVEN ACCESS! That is in addition to the $300 that was already in my ad account...So now I am paying $1049.99 for ads I cannot run, in an account that I don't have access to. I had to contact PayPal and cancel my current credit card due to this. ...I have been a member of Facebook since 2008 and now all of a sudden with no warning, no notice, they are saying my account is fake and I cannot dispute it, though I'm fairly confident that no one hacked my account...The only way to contact them is on Facebook and I cannot even login to contact them. I have written emails with no response, and I only get a help page that is no help."

"This is my story of how Facebook's corrupt ad system disabled my accounts, locked up my prepaid funds, and left me high and dry with no real explanation or support.

"I started a drop shipping business and created a Facebook business page to run ads.... I had to load prepaid funds into my account. For about a month, I tested products, spent money on ads, and everything seemed fine—until boom, my account got disabled out of nowhere.

"Facebook's reasoning? Their bot flagged my ad for 'deceptive or misleading business practices.' No specifics, no examples, just a vague accusation. I reached out to support, and they...suggested I create a new account with prepaid funds while waiting for a manual review on the first one, since that could take weeks...Desperate to keep my business going, I followed their advice.

"I made a new account, loaded it with $500 of my own money, and started running ads again. I spent $100 testing a product, then two hours later—bam—second account disabled. This time, they hit me with a laundry list of nonsense: 'trying to get around our ad review process,' 'activity harmful to the community,' and 'using automations that take away from authentic human activity.' I didn't do ANY of that. No bots, no shady tactics—just legit ads for my drop shipping store.

"Now, I've got prepaid funds locked up in TWO disabled accounts—money I can't access or get back. I went back to support about my first account, hoping for an update on the review they promised. Their response? 'Permanently disabled, nothing you can do.' No explanation of what I supposedly did wrong, no chance to appeal, just a big middle finger. For my second account, I've requested a manual review and have been ghosted for two months. Crickets.

"This isn't just bad customer service—it's predatory. Facebook pushes you to preload money, dangles hope of reactivation, then bans you without evidence or recourse, keeping your funds hostage. It's a trap designed to screw over small businesses while their bots run wild with zero accountability. I'm out hundreds of dollars, countless hours, and my trust in this platform. If you're thinking of running Facebook ads, beware—they'll take your money and run."

Problem with Bots Clicking Ads

"I ran my ad over 10 days…and put in about $20 per day. So far 100% of the clicks and sign ups I've received are bots. How is this legal? I feel like real people aren't even seeing the ad if only bots are filling out my form. It's obvious they are bots by the information they provide. It's so frustrating…If Facebook is taking $100 from me and the ads are essentially being shown to only bots, how many other people and businesses is this happening to? There must be some way this is illegal."

"FACEBOOK AD'S ARE SCAMS. I ran an Ad. For a free camp (for kids). Received 18 responses and 7 of them were bot responses from actual people in my area with real accounts that FB used to respond to my ad...I checked with several of these people and 4 of them said they never sent me a message. I have reported this to my local FBI for criminal referral. THIS IS THEFT that FB is doing."

Running Ads from Scammers

"I've repeatedly reported ads on Facebook as being scams - a 30 second search will reveal that the websites the ads point to are fake copies of the real ones. If I can figure that out, Facebook should certainly be able to. No, they know that they're taking money to run fake scam ads, and they don't care. A couple days later they'll report back that they investigated and there was nothing wrong with the clearly fake scam ad. Trust nothing on Facebook."

These posts are just a sampling of the many complaints about Facebook advertising, and you can find plenty more if you do a search for "Facebook advertising complaints" or put in related terms. These posts point to a series of common frustrations: ad campaigns halted without clear reason, money taken for ads not run, refund requests unanswered, difficulty in getting any response about advertising problems, and more. These complaints align closely with the problems I heard myself from colleagues spending $1000 or more, having ads declined, finding accounts shut down, not getting money refunded, finding bots driving clicks, and experiencing unauthorized transactions.

More Problems for Facebook's Advertising

I recently wrote an article on Facebook's problems with advertising, "*The Problem with Facebook Ads – From Scams to Losing Money.*" I cited quotes from advertisers who reported different types of problems.

Now I want to provide a more general report on the systematic problems that are facing advertisers – from encountering scams by Facebook advertisers to losses of funds from ads that are ineffective or don't run.

Broader trust and platform integrity issues

Beyond advertiser-to-platform frustrations, there is a deeper trust problem:

- Reports show that scam ads continue to run on Facebook and Instagram, sometimes leveraging fake websites or product sales that collect money and never deliver.
- Advertisers not only risk poor performance but are operating in an environment where the ad platform may be prioritizing scale and revenue over full transparency and satisfaction.

The major risks of placing ads

For advertisers (especially small and medium-sized businesses), there are a number of risks in placing ads:

- **Budget risk**: Poor performance, bot clicks, or non-running campaigns mean wasted ad spend.
- **Refund risk**: If the platform declines ads or disables them after funds have been applied, obtaining a refund may be non-straightforward or impossible.

- **Scam risk**: Even as advertisers, businesses may find their brands side-by-side with bogus ads, or even victims of unauthorized transactions or bot-driven click traffic.
- **Trust erosion**: If user confidence in the platform's ad ecosystem drops, the overall return on ad spend may decline, reducing the value of placing an ad.

The many common problems that advertisers encounter

Although many of the specific dollar amounts from $20 to thousands of dollars appear in anecdotal accounts rather than publicly verified case studies, they express common themes::

- Ads are disabled mid-campaign and fund are not refunded.
- Ads generate click traffic but result in little to no real conversions or sales, often because the clicks were bots.
- Unauthorized or unexpected charges appear on accounts tied to Facebook ad spending.

Why these problems persist

While Facebook has claimed it has made some effort to address these problems by canceling hundreds of thousands of allegedly phony accounts, these many problems have continued due to several factors:

- **Scale**: With billions of users and millions of advertisers, monitoring and enforcement are extremely challenging. The internal documents suggest Meta itself views the trade-off between enforcement and ad revenue as significant.
- **Algorithmic complexity**: Ad delivery is increasingly automated and opaque. As one Redditor put it, "This place is flooded with 'ads just stopped working' posts."
- **Fraud ecosystems**: Click fraud, bot farms, and scam advertisers operate in complex ways. Ad fraud, such as due to click farms and fake traffic/ has been documented across platforms.
- **Support and transparency gaps**: Many advertisers report difficulty contacting or receiving meaningful responses from platform support when things go wrong, such as ads decline, budgets overspend, and accounts disabled. They find it impossible to talk to anyone or even receive emails in response to their report of a problem.
- **Platform incentives**: The platform gets paid when advertisers spend. If ads don't run, or run poorly, the platform may still have captured the funds, and advertisers often don't get a refund.

What advertisers and businesses considering advertising should do

Based on this situation, here are some practical tips:

1. **Monitor spend and performance daily**. If you see unusual charges, bots driving clicks, or stalled campaigns, act quickly to report the problem – or even cancel your credit card or other payment arrangements.
2. **Set conservative budgets and short time-frames:** When testing ads, especially on platforms like Facebook where the risk is higher.

3. **Document everything**: Take screenshots of campaign settings, approvals/denials, screenshots of charges. This helps if you need to request a refund or escalate your complaint.
4. **Use tracking and conversion verification**: Don't just watch clicks - ensure conversions are real, monitor bot detection, and look for suspicious traffic patterns.
5. **Be prepared for account issues**: Some advertisers have had accounts disabled with little notice. If your account is new or your business is risky, anticipate this and plan accordingly, such as by arranging for advertising on other platforms.
6. **Have a fallback or diversification strategy**: Relying fully on Facebook ads can expose you to platform risk; consider other advertising channels at the same time.
7. **Stay informed**: Follow communities that report on advertising experience, such as Reddit's Facebook Ads group, and monitor regulatory/industry reporting around scam ads and ad accountability.

The credibility problem

The advertising ecosystem of Facebook/Meta is still massive and influential. But for many advertisers, the experience is increasingly one of frustration, suspicion and risk - from unreturned budgets to bot-driven traffic to scam ads running on the platform. If advertisers and businesses are to continue trusting Facebook with their dollars, the platform must step up transparency, responses to refund requests, and enforcement of advertisement quality. Until then, skepticism and caution are warranted.

Still More Facebook Problems

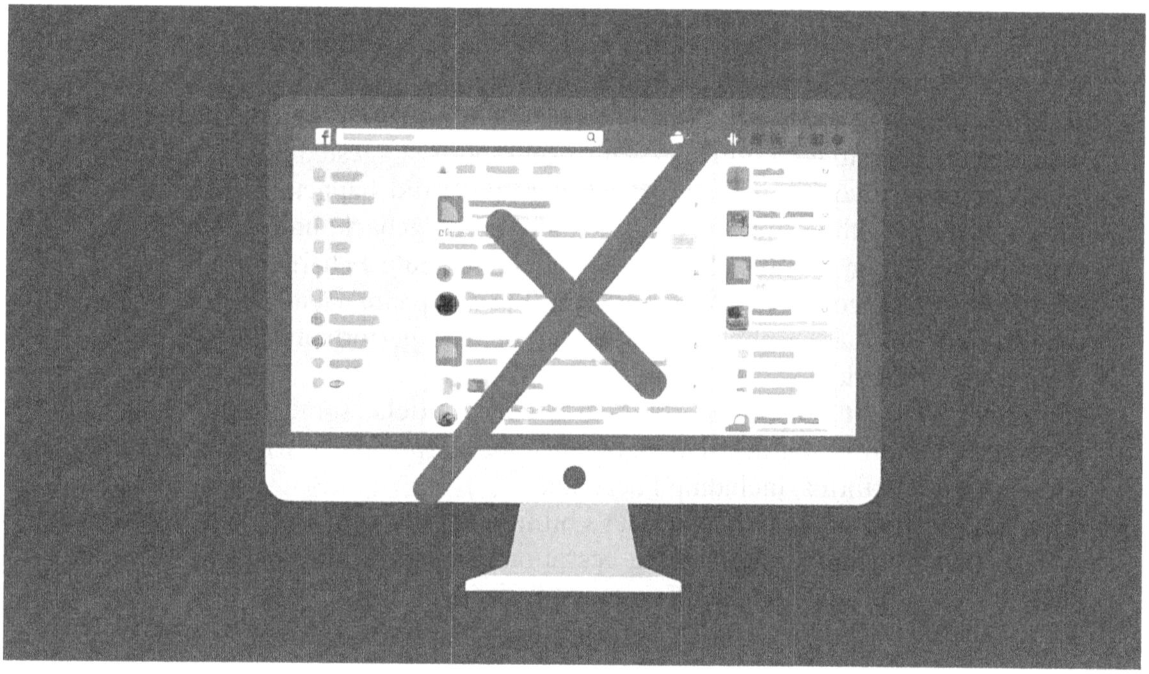

I previously described problems that advertisers had with Facebook ads, the extent to the problem, and some cautions that prospective advertisers might take, including advertising on other platforms. This article discusses these problems and includes some additional examples of the many types of problems advertisers have encountered, including getting bots instead of real readers, getting misleading results about ad performance, and losing money from ads that don't run or are pulled back in the middle of a campaign.

The extent of the problem.

These stats from various Internet sources show the extent of the problems with scams, consumer losses, a failure to get refunds, and incorrect metrics:

- **Scam/abuse at scale:** Internal documents reported by *Reuters* indicate that Meta projected that 10% of its 2024 revenue came from advertisers of scam goods, but the company limited enforcement to avoid reduced revenue hits. This apparently occurred due to Facebook having strict internal guardrails on enforcement to protect its revenue. In other words, it had systemic pressure to let the abuses persist to prevent revenue losses, with the result that Facebook had billions of scam ads shown daily.
- **Scam ads continue to appear:** An article in *The Washington Post* reported there were documented deceptive paid ad placements, such as for tariff relief scams, that slipped through the Facebook ad review process. Researchers also found that deceptive ads

repeatedly made it through these reviews. For example, the researchers found that about 50% of a sampled set violated Meta policies on a given day.

- **Consumer losses keep rising:** The Federal Trade Commission reports that consumers experienced $12.5 billion in fraud losses in 2024; much of this due to investment scams in social-media ads on Facebook and other platforms.
- **Refund timing**: According to a Reddit reporting of Facebook's guidelines, "It usually takes at least one month to get a refund." Many user reports suggested that it took anywhere from 24-48 hours to two to eight weeks when refunds did happen. But many users never received credit without filing a bank dispute or chargeback, and if they waited too long, typically more than 60 days, they were unable to dispute a charge.
- **Metric/integrity concerns:** As reported in the *Financial Times,* a former Meta product manager alleged that Facebook promoted inflated ad-performance metrics, although Meta denied any wrongdoing.
- **An effort to fight back:** According to a *Wired* article, some former Meta advertisers launched a nonprofit to track scam ads, since they experienced widespread trust gaps with social media platforms, including Facebook
- **Charges on stopped ads:** While Meta's billing helps to explain how multiple charges can occur, users frequently note lagging cessation in charges after pausing their ads, so that stopped ads or billing continued for stopped ads.

More Examples of Advertiser Complaints

Following are still more examples of problems that advertisers have experienced. These examples come from ChatGPT50, which has found these cases through internet research They reflect four common themes:

- **Refund failures** are common even when the ads never ran.
- **Unauthorized charges** often require banks or card issuers to intervene to stop them
- **Bots and fake clicks** waste ad budgets.
- **Support is slow, inconsistent, and often unhelpful.**

Charged but Account Disabled Immediately

A user funded a new ad account with €24, saw the charge cleared by their bank, but Facebook's dashboard flagged it as "payment failed." The account was instantly disabled, leaving them unable to access ads or funds. As the user complained: "My card was charged, but the ad account was disabled. Support told me to wait 48 hours — it's been weeks."

Payment Method Rejected and Account Locked

An advertiser added a valid card and PayPal, but every attempt triggered "We couldn't charge your payment method." The system froze the account before any ads could launch. According to the advertiser: "Meta kept saying my payment failed — then disabled the account. Now the billing section's gone completely."

Unauthorized Ad Charges on Credit Card

A Reddit user reported recurring Facebook ad transactions on a card never linked to any Facebook account. Their bank confirmed the charges were processed by Meta Payments, leading them to feel frustration over Facebook's fraud screening. As the user complained, "These are clearly fraudulent Facebook ad charges. How is this even allowed?"

$0.01 Glitch Disables High-Spending Account

One business that spent $1,150 each day found its entire campaign disabled over a $0.01 unpaid balance. Even after paying the penny, the account remained frozen. According to the user, "It's absurd — I'm spending over a thousand a day and Meta shut me down for one cent."

Account Hold and Payment Loop

Several advertisers repeatedly report being trapped in a "temporary hold" status where no payment method is accepted. Debit cards, credit cards, even PayPal fail, and Meta's automated support offers no solution. As one of the advertisers complained, "Every payment attempt fails. Meta says wait 24 hours, but it's been weeks — business is at a standstill."

Ex-Client Spends $20,000 Using Old Agency Card

An agency discovered a former client had left their card on file and used it to buy about $20,000 in Facebook ads. The agency faced the choice of disputing the charges — risking a permanent Facebook Ads ban — or absorbing the loss. The ad agency reported that ""Meta won't even talk to me about the fraud because the account technically belongs to the client."

Ad Account Disabled Mid-Campaign

A small business owner running new Shopify ads was hit with a generic "unusual activity" notice and account suspension, stopping all ads and wiping saved audiences. According to the business owner, "I finally had a campaign that worked — now Facebook says I can't advertise anymore with no explanation."

Charged Daily After Ads Blocked

One advertiser ran a single ad, but months later their card began getting charged daily even though no active campaigns existed. Refund requests went unanswered. Eventually, the advertiser reported that: "I had to cancel my card. Meta kept taking money even after everything was stopped."

Disabled Account Still Billed for Ads

An advertiser whose account was disabled continued to receive invoices for prior impressions and even for new activity. Meta claimed these were delayed settlements for earlier ad delivery, but the advertiser still received invoices. He complained that "They disabled me for policy reasons but still billed me after the ban. Feels like theft."

Verification Loop Prevents New Ads

Some advertisers found themselves caught in a loop after they added funds to a new account, when Meta's system entered an endless verification loop, which locked them out of the campaign dashboard. They found the support bots unhelpful, since they gave only generic responses. As one advertiser described the situation: "Every time I try to verify, it says 'error, try again later.' I'm basically paying for nothing."

Bot-Driven Traffic and Zero Results

Numerous advertisers report high click rates but no engagement — evidence of a bot or click-farm traffic. According to one of these advertisers: "ABSOLUTELY 100% of the traffic are bots. Thousands of clicks, zero site visits." Other advertisers described identical behavior when they placed ads.

Budget Limits Ignored or Exceeded

Though advertisers set daily caps, such as $50/day, they were billed $70 or more, sometimes due to Meta's flexible delivery algorithm. As one advertiser complained: "I had a $50 daily limit but got charged $76. Support told me it's within 'allowed variance' , but that's not okay."

Ads Canceled but Billing Resumed Months Later

Several small-business owners say Facebook reactivated canceled ads, charging hundreds of dollars without consent. According to one advertiser, "I stopped my ad in April. In July it restarted by itself, and I got billed $600."

Disabled for "Misleading Practices," No Refund

A new advertiser pre-loaded money for an ad credit, but before running a single ad, Meta flagged the account for "deceptive or misleading business practices." Despite no delivery, Meta kept his funds. As the advertiser reported, "They disabled the account for a policy violation before the ad ever ran — and still took the money."

Inactive Page, Sudden Unauthorized Transactions

One user with an inactive page woke up to 20 unauthorized ad charges overnight. The billing history showed campaigns they never created. As the user complained, "My account's been inactive for months — suddenly, $200 in charges appeared. When I tried to report the problem, Meta support ghosted me."

Multiple Small Charges Exceeding Budget

An advertiser set a $5 daily budget but saw repeated micro-charges of $2 each totaling $24 with little performance data. According to the advertiser, "I can't even track where these tiny charges came from — they just keep stacking up."

$0.01 Bug Disables Major Business Account (Variant)

A second major advertiser was locked out of Business Manager due to a $0.01 payment error. The glitch disabled all connected client accounts. The advertiser reported: "Meta support said 'just pay the balance' but it's already paid. We lost a week of client campaigns."

Chargebacks Risk Permanent Ban

Some advertisers resort to chargebacks when Meta fails to pay a refund, but doing so risks losing access to all Meta ad platforms permanently. According to one of these advertisers:
"We filed a chargeback for $750 after months of silence; the next day, our entire ad account was banned."

Practical safeguards to take

Some strategies to avoid these problems, in addition to those already suggested in my previous article, are these:

1. **Daily monitoring:** Check your spending on ads and the conversion to sales to see if your advertising budget is worth the results. Also, investigate mismatches, such as getting thousands of clicks while you have near-zero conversions.
2. **Tight tests:** Start with a short time to advertise to test the results, such as a month to start, and avoid going after ultra-cheap costs per click, if it results in low-quality bot traffic.
3. **For disputes:** Use Meta Billing > Billing Dispute and document everything. The way it is supposed to work is that you log in to your account, go to Settings and Privacy, then to Meta Pay, Activity, select the transaction, click Get Help and Contact Us, and choose Live Chat or Email Meta to report the unauthorized charge as soon as possible. Also contact your bank to report fraud and block the card if necessary. Yet despite these guidelines, in practice, advertisers report a difficulty of hearing back from anyone when they try to report a problem. So perhaps the best solution is to use credit card or bank chargebacks to stop further withdrawals and force a review.
4. **Account protections:** Before you advertise, set up a strong Two Factor Analysis system, have limited ad-account roles, establish low spending limits, set up alerts for unusual activity; and consider separate cards or bank accounts for ad spends, so you can more easily monitor and stop any overcharges.

The Facebook Ad Crisis: How Advertisers Are Losing Money, Trust, and Patience

This article features the highlights of three previous articles I wrote about the problems with Facebook advertising.

A Platform Losing Credibility

Once the most powerful advertising platform in the world, Facebook is now facing a growing crisis of trust. While its executives praise the precision of its targeting and the reach of its 3 billion user network, thousands of advertisers are reporting a very different experience - one filled with bot clicks, unauthorized charges, and missing refunds.

A November 2025 *Reuters* investigation revealed that Meta earns an estimated $3.5 billion every six months from what its own staff classifies as "higher-legal-risk" or potentially scam-linked ads. Internal documents also show Meta's policy teams were barred from taking enforcement actions that might reduce revenue by more than 0.15% or approximately $135 million.

In short: the company knows some ads are fraudulent, but enforcement is limited by profit concerns. At the same time, the company has problems with cancelling ad accounts or not running ads while charging advertisers money and overcharging for ads beyond the advertisers' budgets. Then when advertisers complain and try to report the problem, they find it difficult or impossible to reach anyone to resolve their complaint. Thus, in the face of all these advertising problems, legitimate advertisers, who are primarily small businesses, creators, and agencies, are paying the price.

The Rising Cost of "Invisible" Fraud

Facebook's advertising system relies on automation: algorithms approve or reject ads in seconds. That speed comes at a cost. A growing share of advertisers report that campaigns are wrongly flagged, disabled mid-run, or attract bot-driven fake clicks that exhaust their budgets with no real results.

A Reddit thread titled *"Meta Ads Click Fraud Rate for 2024"* reported that over 22% of "problematic" Facebook ads had no direct human targeting, meaning that Facebook's own algorithms were placing them. At the same time, global consumer data show that social-media ad scams are rising sharply. The FTC reported $12.5 billion in total fraud losses in 2024, with nearly one in four victims saying the scam began with a paid social ad. A *Financial Times* report in 2025 alleged that Meta inflated its ad performance metrics to maintain its targeting power, while *Wired* documented internal whistleblowers launching nonprofits to track scam ads after Meta failed to remove them.

In dozens of reported complaints, the same pattern emerges:

- **Unreturned funds:** Advertisers often lose prepaid balances or payments for ads that never ran.
- **Unverified fraud:** Unauthorized ad charges can linger for weeks without clear resolution.
- **Bot traffic:** High click rates and low conversions suggest extensive fake engagement.
- **Unhelpful support:** Meta's automated help center offers generic replies and rarely issues refunds.

A Breakdown in Trust

These problems are systemic, not just an occasional glitch, due to an incentive to maintain earnings. When Meta earns billions from automated ad spend, it has little motivation to overhaul a system that generates revenue even when the ads misfire. As one former Facebook ads manager told *Wired*: "The platform treats small advertisers as disposable. There's almost no accountability once the payment goes through."

What Advertisers Can Do

1. **Track daily performance.** Monitor your spending, clicks, and conversions closely.
2. **Start small.** Limit your budget and test your campaign's performance before scaling up.
3. **Document everything.** Keep screenshots of transactions and ad approvals.
4. **Use credit card and bank chargebacks as last resort.** While these efforts may recover funds, they but could trigger account bans.
5. **Diversify ad channels.** Use TikTok, YouTube, Google Ads, or other social media platforms to reduce your platform dependency.

The Bottom Line

Facebook's ad platform remains vast, but its reputation among small and mid-size advertisers is eroding fast. Between refund failures, bot-driven fraud, and indifferent support, many now see Meta as a high-risk partner rather than a reliable marketing channel.

Unless the company invests in real transparency, faster dispute resolution, and stronger fraud detection, the world's biggest social network may find its advertising engine losing its advertising appeal, due to a decline in trust and an increase in ad failure and frustration.

Facebook's Problem of Too Many Ads

I recently wrote a series of articles on the problems with using Facebook and published them in a book *WHAT'S WRONG WITH FACEBOOK: THE GROWING PROBLEMS OF FACEBOOK WITH SELFIES, SCAMS, AD FAILURES AND FRAUDS.* I thought I had covered all the major problems, but when I talked about my book with my banker, I discovered another serious problem I left out – too many ads in their feed that is turning off users and leading many to leave the platform. And if users are angry about seeing too many ads, that's also a problem for advertisers, since not only can users leaving reduce their target market for viewing their ad, but many users are trying to hide or avoid seeing their ads.

I realized that seeing too many ads was a problem for Facebook and its users when my banker told me:

> "I'm about to leave Facebook because I see so many ads. I'm hoping to get messages from my friends, but then I have to plow through so many ads to see them. It's very frustrating, so I'm ready to leave – and some of my friends will too because they don't like all the ads either."

The reason for all these ads is that advertising enables Facebook to provide a free platform for users. It makes its money from ads rather than charging users to become members or subscribers. However, Facebook seems to be posting too many ads, which is undermining the user experience and leaving users feeling angry and even trapped, because they have been on the platform for so long and have a network of friends and group activities there. The result is that a growing number of users have been turned off by too many ads. So many may try to ignore the ads or leave the platform, even if they are a long term user with a large friends' network, because they can't stand wading through so many ads.

Examples of Complaints about Ads

Here's a sampling of the many postings I saw online from Facebook users complaining about the ads. Mostly the complaints come from Google search or Reddit, who griped about too many ads overwhelming the information they got about friends and family, or getting ad that had nothing to do with their interests. And some felt they were seeing more ads than usual in the past year, when Facebook began featuring many more ads. As a result, some were ready to leave Facebook for some other social media platform. Here's what they said:

> "For me, big change started about 1 year ago. My feed now flooded with pages totally unrelated to my interests, almost all of them counterfeit pages of 'celebrities' and TV shows."

> "The whole point of Facebook is keeping with friends and relatives. It used to be that way: now they're dumping the things I want to see and showing me all kinds of stuff I don't want to see."

> "My feed changed in the last couple weeks. It's honestly insufferable now. Just random pages and memes that I don't follow. I absolutely hate it and spend way less time on that app."

> "For me, blocking unwanted shit only made me seeing more and more of said shit, just posted by different scam accounts. Roughly 50% of my feed is groups I'm not member of, advertisements, sponsored posts and for the last 2 months – posts from profiles I don't follow, never heard of before, and they aren't even flagged as sponsored."

> "If I'm in the home tab (mobile), I get ads every 3 or 4 items and non-stop influencer/porn adjacent crap and rage bait. I've been blocking and clicking 'suggest less' for weeks with no perceptible difference."

> "I just hit/snoozed/deleted/reported and as soon as I got rid of one ad page and refreshed, the next would pop up. It felt like it was groundhog day."

> "Now I am getting clickbait and low quality content every few posts on my feed. No matter how many times I choose to 'Hide all from xyz,; I just get more of it. It's unusable and the good info I am getting from the platform isn't worth wading through the shit."

> "What has Facebook become? 99% of my feed is ads and irrelevant garbage groups."

"What is going on with all the ads? Opened Facebook today and now my feed is like 80% ads."

"My Facebook feed is nonstop trash posts I don't want to see. Is there any way to stop it."

"Are they trying to kill Facebook? Feeds are ridiculous now. Just ads, groups, and trash content."

"When is Facebook going to quit forcing so many ads into our feeds? I see more ads than friends' posts and by this notification I just received, I don't think it'll stop anytime soon... Greediness is real."

"I'm now getting 42 or more ads between actual content on my Facebook feed!"

"How can I block unwanted posts from polluting my feed? I'm tired of hiding them."

"Facebook is only ads and not content I actually follow anymore! How do we fix this??"

And then I saw this – a detailed post about all the problems wrong with Facebook's increase in advertising.

"In the last year or two it's gotten worse. I really think Facebook is making terrible decisions, and their algorithm is going to drive people off the platform. My feed is just full of ads, sometimes multiple ads in a row, and rarely about anything I'm interested in…

"For years, my purpose in looking at my Facebook feed was primarily to find out how my friends were doing, plus to see posts from groups that I had specifically chosen to join. But now they've really screwed it. The experience is no longer fun and is just frustrating and enraging. I scroll through and get angry constantly at the crap they're throwing into my face instead of seeing my friends' baby or travel photos, etc.

"So what's my incentive to even look at my Facebook feed anymore, or even to use Facebook? If they're going to make the experience of my feed into something that mostly has content I don't care about, they're really shooting themselves in the foot. And I think this will lead to their demise. I believe most people are on Facebook to connect with their close friends and family, and secondarily for groups that they are specifically interested in. Facebook's algorithm is trying way too hard to get engagement on posts and pages that they think the person might like.

"…I'm just not on Facebook to have random content shoved in my face that I have to constantly hide or reject. It's really annoying, and definitely more than half of my feed now is just ads and suggested posts.

"…is this just the new normal? And what do people think about the future of Facebook, given how they've made the feed experience into such dogshit?.. At least they could offer the option to pay extra to customize your feed and remove ads. I would actually pay it. But they don't give such an option."

Why So Many Ads

The reason users are seeing so many ads is because that's how Facebook makes money, since it doesn't charge for using the platform. Moreover, you are supposed to see more ads related to your interests, since Facebook gathers data about its users, so it keeps track of when you like any posts, watch videos, or follow a person or brand. It even tracks you off the app, such as when you look up a product on Amazon or another website, since Facebook can place a pixel on other sites to show what you viewed or searched for, even when you aren't using the Facebook ap.

By getting information from other sites, Facebook does what advertisers call "retargeting." That allows advertisers to retarget people who have visited the company's website or app on Facebook. Thus, if you click "like" on a product or page, Facebook can use that information along with your name and profile picture to show targeted social ads on your friends' pages, and third-party advertisers can use Facebook to promote their product. However, targeting an audience using your personal information from your profile raises privacy issues, which can also be a concern, since you haven't given permission to others to use your information. But that's another concern apart from the onslaught of advertising.

As described in an article by Jennier Wilson, "Why Is My Facebook Feed All Ads," Facebook changed its algorithm about seven months ago, so now it prefers to show you paid content, so "regular posts from friends and family get pushed down. On top of that, the sponsored ads take center stage." The reason for all these ads is that "Facebook gets paid when you click on ads. So it wants to show you ads that you'll likely click" And that shift in its algorithm means that "your news feed is now full of…sponsored posts…suggested reels from brands…and product ads that match your browsing history."

But while these tailored ads may help to target more users with ads they will click, this approach is alienating many users, who feel overwhelmed by all the ads, find them irrelevant and annoying, and may want to leave Facebook for other social media platforms.

What to Do About It

While you can't get rid of the ads entirely, you can reduce them or possibly see ads that are more aligned with your interests. There are several ways to do this:

- Switch to the Friends feed. Though you'll still see some ads, they'll be outnumbered by the feeds from friends.
- You can get a Facebook Purity extension for your browser on your computer, though it won't work on a phone or tablet. The extension can work on Chrome, Firefox, Edge, Safara, and Opera browsers to remove all apps from your feed.

- Hide certain ads and advertisers on a computer by clicking or tapping the three dots on the ad and select "Hide."

And if you are still frustrated by the ads you keep seeing on Facebook, you can always leave for another social media platform.

What's Wrong with Facebook: A New Privacy Settlement and a New Book on Facebook's Problems

Facebook has evolved from a friendly social network into a sprawling, automated system where ordinary users, small businesses, and even nonprofits can lose access to their accounts overnight, while scammers and fraudulent advertisers often operate with shocking freedom and little accountability. Recent privacy settlements and lawsuits show that these are not isolated glitches but symptoms of a deeper structural problem: Facebook's business model and governance consistently prioritize data extraction and ad revenue over user safety, privacy, and due process.

Facebook's recent privacy settlement and a new book: *What's Wrong with Facebook: The Growing Problems of Facebook with Selfies, Scams, Ad Failures, and Frauds* provide a harsh look at the problems with Facebook and what to do about it.

A platform built on privacy failures

Over the past several years, Facebook's most visible "fixes" have come only after public scandal and intense legal pressure, not proactive responsibility. The landmark 725 million dollar consumer privacy settlement — sparked by years of litigation over Facebook sharing user data with third parties like Cambridge Analytica — was granted final approval in 2023 and upheld on appeal in early 2025, making it the largest recovery ever in a data-privacy class action and the largest amount Facebook has ever paid to resolve a private class action. Millions of users are now receiving small payouts, but the underlying story is stark: for more than a decade, Facebook allowed extensive data harvesting and profiling while insisting users were in control of their information.

This case is not an outlier. Facebook has repeatedly been accused by regulators and plaintiffs of violating its own promises to protect user privacy and comply with earlier FTC orders, leading to a patchwork of settlements and consent decrees that still have not fully stopped questionable data-sharing and tracking practices.

Selfie "verification" and the human cost of lockouts

One of the most disturbing recent developments is Facebook's growing reliance on automated "video selfie" verification systems that can lock people out of their accounts with little explanation and almost no meaningful path to appeal. In theory, these tools are supposed to fight scams and restore compromised accounts; in practice, many legitimate users report that their video selfies are rejected, their accounts remain suspended, and their attempts to reach support lead to silence.

Users describe being prompted to upload a facial recognition style video selfie after a routine login or after creating a new account, only to be told that the verification failed and their profile is under review indefinitely. Some report losing years of photos, contacts, and business pages when the system decides — often wrongly — that they are underage, a bot, or otherwise "inauthentic," with multiple community posts noting that once a selfie is rejected, it is extremely rare to regain access without insider help or third-party services, so most users who become losers can't regain their account.

For ordinary users, this means losing a digital scrapbook of their lives. For small businesses, nonprofits, and creators, it can mean losing customer lists, ongoing ad campaigns, and brand pages built over a decade — erased in an instant by an opaque biometric gatekeeper with no clear human oversight or due process.

Scam bots, fake ads, and the fraud economy

While legitimate users and advertisers struggle with lockouts, scammers and fraudsters often find Facebook surprisingly hospitable. Class-action suits from users who were tricked by fraudulent ads describe a familiar pattern: fake e-commerce shops, impersonated brands, bogus investment schemes, and deep-discount offers that never ship. Yet these ads are routinely approved and heavily promoted.

Internal documents obtained by reporters show that Meta projected around 10% of its 2024 revenue could come from ads for scams and banned goods, suggesting that the company knowingly benefited from fraudulent advertising that harmed its own users. External commentary on those documents notes that Meta sometimes allowed "high-value" advertiser accounts to rack up hundreds of violations without being shut down, effectively tolerating serial scammers because they were profitable.

In lawsuits filed by victims, Facebook users argue that the company broke explicit promises in its terms of service and community standards to "take appropriate action" against harmful and fraudulent content. But instead, it didn't, so that a federal judge found that, if proven, those allegations could support claims that Meta breached its contractual duties and duty of good faith. Meanwhile, some media coverage highlights how these scam ads mimic trusted brands and trick people into clicking dangerous links or handing over credit card details, turning the platform into a high-risk environment for anyone who assumes ads are vetted.

Broken appeals, opaque bans, and advertiser losses

Another core problem lies in how Facebook enforces its rules and how little recourse users have when automated systems get things wrong. Reports from business owners and group admins show accounts flagged or disabled for vague reasons like "account integrity" or "violations of community standards," sometimes for posting ordinary content or even offering free items in local groups.

These users often find that:

- Automated systems disable accounts with little explanation.
- Appeal links lead to canned responses or no reply at all.
- There is no clear way to escalate to a human decisionmaker.

In business and advertising contexts, the stakes are higher. Once an ad account is disabled — sometimes after a hack, sometimes after unexplained "policy violations" — advertisers report losing access to ongoing campaigns and outstanding balances, with limited success getting refunds or restoring access despite years of spending on the platform. For many small businesses, Facebook and Instagram are their primary marketing channels, so a sudden lockout can mean losing a major share of revenue and customer contact overnight, with no equivalent way to reach their audiences elsewhere.

What needs to change

The pattern across privacy settlements, selfie lockouts, scam ads, and failed appeals points to a single underlying problem: Facebook has built an infrastructure where automation and revenue optimization dominate, and human accountability is an afterthought. Users are asked to trust black box systems with their identities, data, and livelihoods, but when those systems fail, the burden falls entirely on the individual—even though they have almost no information, rights, or remedies.

Several concrete changes are urgently needed:

- Stronger enforcement of privacy and consumer-protection laws, with real penalties when platforms violate their own promises or enable large-scale fraud.

- Clear due-process rights for users and advertisers, including timely human review, transparent explanations for bans, and meaningful appeal mechanisms.
- Limits on automated biometric checks — like video selfies — without robust accuracy standards, independent audits, and strict rules against repurposing biometric data.
- Greater liability for hosting and profiting from scam ads, especially when internal documents show prior knowledge and financial dependence on that revenue.

Until regulators set these guardrails and Facebook redesigns its systems around safety and fairness — not just engagement and ad spend — users and advertisers will remain exposed. The stories behind the lawsuits and complaints make one thing clear: what is wrong with Facebook is not just a series of bugs, but a business model that treats human lives, memories, and livelihoods as disposable collateral in the pursuit of growth.

PRESS RELEASES ON WHAT'S WRONG WITH FACEBOOK

Article on Substack: Facebook's Face-Off: Facebook's New Video Selfie Check Leaves Many Legitimate Users Locked Out

Meta's recent implementation of a video-selfie identity check on Facebook is drawing widespread user backlash, as described in a recent Substack article. While intended to accelerate account recovery and block impersonation, the system appears to be locking out countless legitimate users with little recourse.

Facebook's official help pages explain that users may be asked to record a short selfie video, moving their head in multiple directions, so Meta can confirm identity and detect "fake videos." But the real-world user experience results in many people being locked out and losing their accounts.

Long-time users report being funneled into endless loops of selfie resubmissions—with no alternative option to submit a government ID, and no direct access to human support.

In my own case, my account was flagged after inviting a friend; soon afterwards my account was suspended and the only option presented was the selfie check. Even after I tried creating a new account and using a different computer, the system invoked the same verification trigger. User communities are full of similar anecdotes:

The article describes the extent of the problem and includes dozens of complaints about the system.

Compounding frustrations, security experts point out that while biometric checks may raise the barrier for fraud, they are far from fool-proof. Fraudsters are already deploying tools to spoof selfie checks, while real users with older devices, poor lighting, or accessibility constraints often fail the verification process.

Meta says the selfie videos are encrypted, and facial data used only for matching purposes. Nonetheless, the lack of transparency about failure rates, the absence of reliable human support, and the one-size-fits-all automation raise concerns among digital-rights advocates and everyday users alike.

Until Meta offers clear human escalation paths, better fallback options, and transparency about how the system flags accounts, the video-selfie verification remains a precarious experience for many.

You can read the full article on Substack at https://gini.substack.com/p/facebooks-face-off-how-facebooks.

Article on Substack: Facebook's Decline and Video Selfie Problems

As the article on Substack: "Facebook's Decline and Video Selfie Problems" describes, Facebook is in a state of decline, characterized by a loss of daily users and ad sales. Contributing to that decline is its "video selfie" identity program. Not only is Facebook suffering from decreased engagement, less interest from a younger audience, and lower ad revenues, but a growing number of users have dropped out because of Facebook requiring a video selfie to prove one is a real person. The big problem is that the system is locking out real people since the picture taking feature doesn't work very well. That even happened to me after I had an account for over 20 years, and I was unable to get it back despite multiple attempts to take my picture.

The article points out that this loss of users is about half a million in recent months. While Facebook has about 2.11 billion daily active users and around 3.07 billion active users each month, it has recently lost about 500,000 daily users according to the *Washington Post*. And Facebook's selfie-program is a major source of this loss of users affecting millions of people. A U.S. local tech segment reported that "some 110 million people" get locked out of Facebook accounts annually, many due to selfies.

The way this program works is you have to prove your identity by recording a short selfie in a circle while you turn your head up, right, or left according to a series of random arrows. Then, Facebook uploads a file with your image, sees if it meets its criteria for an acceptable image, and if not, you may be asked to retake it or receive a decline.

Supposedly, Meta set up this system to avoid people setting up fake accounts, curb scams, deepfakes, large-scale impersonation campaigns, and fake IDs. But many real people can't pass the test. Even people with long-time accounts may suddenly find they have to prove they are a real person due to using their account in a way that draws Facebook's attention. Then, they have to go through this selfie process, and if they can't pass the test, they lose their account, may find it difficult or impossible to appeal the decision, and can't speak to anyone personally to restore their account.

The article describes Facebook's decline and problems with its video selfie program in more depth. It discusses my own experience and that of dozens of other former Facebook users who lost their accounts.

You can read the full article on Substack at https://gini.substack.com/p/facebooks-decline-and-its-video-selfie.

New Book Exposes What's Wrong with Facebook — From Rampant Hacks and Broken Identity Checks to Ad-Scams Costing Users Thousands

A new investigative book pulls back the curtain on Facebook's deepening crisis of user trust, platform failures, identity-verification breakdowns, and widespread advertising scams. *WHAT'S WRONG WITH FACEBOOK? THE GROWING PROBLEMS OF FACEBOOK WITH SELFIES, SCAMS, AD FAILURES, AND FRAUDS* published by American Leadership Books and written by prolific author and social-trends analyst Gini Graham Scott, PhD, JD, is a timely and comprehensive examination of a social network once considered indispensable—and now widely regarded as unreliable, unsafe, and increasingly irrelevant. A video introducing the book is featured at https://youtu.be/fQWD1y2rPOU.

As Facebook has expanded automated enforcement, outsourced moderation, and algorithmic identity checks, millions of users have reported losing access to their accounts without warning or recourse. Meanwhile, hackers, impersonators, and scammers have flourished, exploiting weaknesses that Facebook has failed to meaningfully address. Scott's new book outlines these systemic problems with clarity, accessible storytelling, and practical step-by-step guidance for protecting online identities.

A Four-Part Breakdown of a Platform in Peril

The book is structured in four sections, each revealing a dimension of the company's growing instability:

• Problems with Hacks

Scott opens with the widespread surge in hacked accounts, including her own experience, and details how both casual criminals and organized rings exploit Facebook's vulnerabilities.

Chapters explore why hackers target accounts, how they break in, and what users can realistically do to recover access despite Facebook's slow or nonexistent support systems.

• **Problems with Selfies**

A deep dive into Facebook's much-criticized "video selfie" identity system shows why many thousands and even millions of users fail verification even when following instructions precisely. Scott documents the widespread complaints, the platform's declining user base, and the frustration of being locked out indefinitely due to software errors that Facebook refuses to acknowledge.

• **Problems with Advertising**

Drawing from real cases, business owners, and consumer reports, Scott reveals a troubling pattern: advertisers losing hundreds or thousands of dollars to fraudulent clicks, unauthorized charges, nonexistent customer service, ads rejected for reasons that Facebook's own systems cannot explain, and too many ads interfering with user's experience. A series of chapters, *The Problem with Facebook Ads*, *More Problems for Facebook's Advertising*, *Still More Facebook Advertising Problems, A Problem with Too Many Ads* showcase a platform in an escalating crisis of transparency and trust.

• **The Latest Problems**

The book concludes with Facebook's most recent failures—from malfunctioning systems to sweeping automated bans that leave users locked out with no human appeal. Scott illustrates how the company's rigid, opaque processes leave ordinary users vulnerable while scammers continue to thrive.

A Guidebook for Users Who Want Answers and Protection

While *WHAT'S WRONG WITH FACEBOOK?* exposes critical issues inside one of the world's+ largest social platforms, it also empowers readers with solutions. The chapters *Recovering Your Account* and *Protecting Your Accounts in the Future* offer practical actions, checklists, and prevention strategies to minimize risk and find alternate social media platforms as users and advertisers due to the problems they have experienced in using Facebook.

A Must-Read for Consumers, Businesses, and Policy Leaders

Whether you've lost your account, wasted money on ads that never delivered, struggled with Facebook's selfie-verification maze, or simply watched the platform decline, this book offers insight into what is going wrong—and what users can do next. As Facebook continues its uncertain evolution, *WHAT'S WRONG WITH FACEBOOK?* arrives as an essential guide to navigating a system increasingly experiencing problems for both users and advertisers.

From Fines to Subpoenas: New Article Calls for Tough Federal and State Action to Make Facebook Fix Its Platform

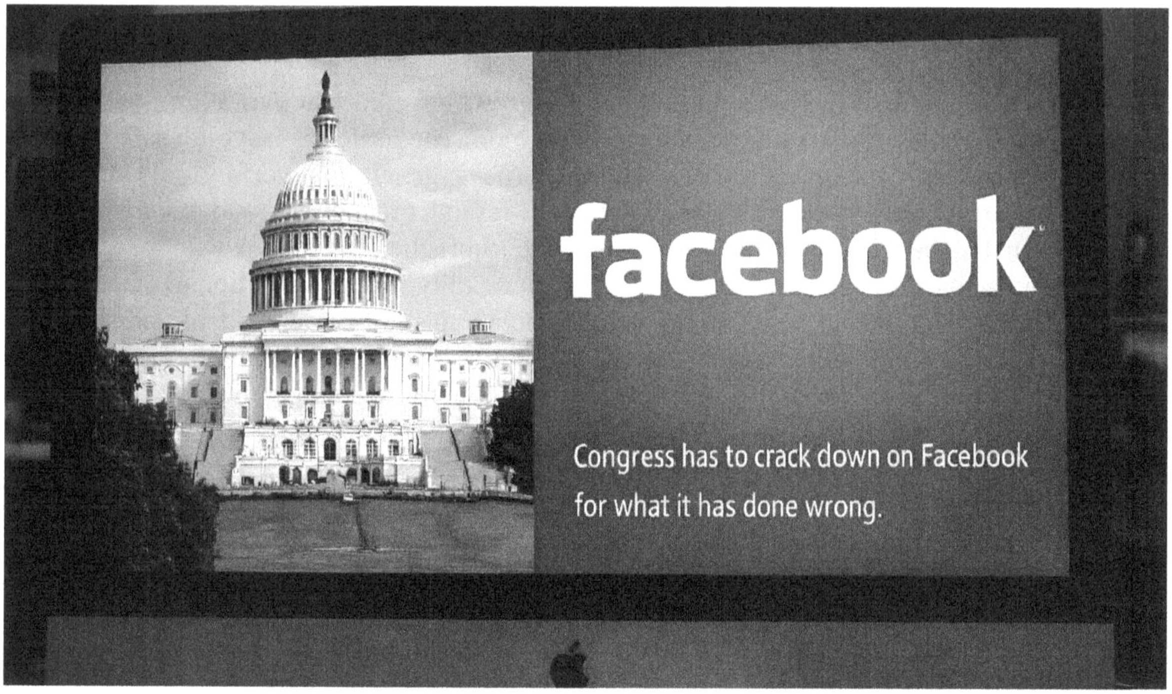

A new article published on Substack and Medium, titled *"From Fines to Subpoenas: What Washington and the States Must Do About Facebook,"* calls on federal and state officials to stop treating Facebook's misconduct as a public relations problem and start treating it as a systemic governance failure that demands serious penalties and structural reform.

The article argues that years of privacy violations, account lockouts tied to flawed selfie verification, rampant scams, bot activity targeting ads, and unfair treatment of advertisers show that Facebook's internal incentives are fundamentally misaligned with user safety and fairness. It contends that regulators have relied too heavily on promises of self-regulation, allowing the platform to treat massive harms as a cost of doing business. The complete article is at https://gini.substack.com/p/from-fines-to-subpoenas-what-washington on Substack.

Focusing heavily on what government can and must do now, the piece lays out a concrete roadmap for action in Washington. It urges the Federal Trade Commission to turn its existing orders and past fines into real leverage—using escalating penalties, stronger compliance obligations, and meaningful executive accountability when Facebook violates privacy and consumer protection rules. It also calls on Congress to use hearings and subpoenas to force disclosure of internal data on scams, bots, wrongful account suspensions, and advertiser disputes, building a factual record that supports tougher legislation.

The article places particular emphasis on the role of key House and Senate committees that oversee commerce, consumer protection, and antitrust. It proposes that these committees aggressively investigate Facebook's handling of fraud and account enforcement, question executives under oath, and consider new laws requiring clearer due process for users and advertisers, stronger privacy protections, and a duty of care to prevent foreseeable scams on the platform.

At the same time, the piece highlights the growing importance of state governments in regulating big tech. It argues that state privacy laws, biometric protections, youth safety measures, and broad consumer protection statutes give attorneys general powerful tools to challenge deceptive practices, demand better identity and ad-refund procedures, and secure meaningful settlements that change how Facebook operates in their jurisdictions.

With Facebook's reputation eroding and users steadily shifting attention to other platforms, the article warns that this is a critical window for policymakers. Either federal and state leaders will seize the moment to force deep reforms—through fines, subpoenas, binding orders, and, if necessary, structural remedies—or Facebook will continue to externalize the costs of its business model onto users, small advertisers, and the broader public.

"From Fines to Subpoenas: What Washington and the States Must Do About Facebook" is available now on Substack and Medium.

For more information, *What's Wrong with Facebook?* is available on Amazon https://www.amazon.com/Whats-Wrong-Facebook-Problems-Failures/dp/B0G3H481RD. To set up interviews or speaking engagements with the author, Gini Graham Scott, PhD, contact:

ABOUT THE AUTHOR

GINI GRAHAM SCOTT, Ph.D., J.D., is a nationally known writer, consultant, speaker, and seminar leader, specializing in business and work relationships, professional and personal development, social trends, science, and crime. She has published over 50 books with major publishers. She has worked as a ghostwriter with dozens of clients on books and proposals on popular business, self-help, and memoirs. Additionally, she writes film scripts and has produced 19 feature films, documentaries, and TV series, with three on scams – *Conned: A True Story, Con Artists Unveiled,* and *Scams in the Digital Age.* Three books on scams are *The Big Con, I Was Scammed,* and *Scams in the Digital Age.*

She is the founder of Changemakers Publishing, featuring books on business, psychology, self-help, and social trends. The company has published over 200 print and e-books and over 170 audiobooks. She has licensed several dozen books for foreign sales, including the UK, Russia, Korea, Spain, and Japan.

She has received national media exposure for her books, including appearances on *Good Morning America, Oprah,* and *CNN.* Her latest books include *The New American Middle Ages* and *Turn Your Book or Script into a Film*, published by Waterside Productions. Her other books on scams and investigations include:

- *Scam Story* published by American Leadership Books
- *Scammed: Learn from the Biggest Consumer and Money Frauds How Not to Be a Victim,* published by Allworth Press,
- *Lies and Liars: How and Why Sociopaths Lie and How You Can Detect and Deal with Them,* published by Skyhorse Publishing,
- *Preventing Credit Card Fraud: A Complete Guide for Everyone from Merchants to Consumers* with Jen Grondahl Lee, published by Rowland & Littlefield,
- *Internet Book Piracy: The Fight to Protect Authors, Publishers, and Our Culture* published by Allworth Press.

In addition, she has half a dozen children's picture books published by various companies. Her other films in distribution include *Driver, Conned: A True Story, Con Artists Unveiled,* and *AI Evolution* (Gravitas Ventures), *The New Age of Aging* and *Courage to Continue* (Factory Films) and *Me, My Dog and I* and *Rescue Me* (Random Media). Details: www.changemakersproductionsfilms.com.

Scott is active in several community and business groups, including the Danville and San Ramon Chambers of Commerce. She does workshops and seminars on self-publishing and using AI for writing and illustration.

She received her PhD in Sociology from the University of California, Berkeley; her JD from the University of San Francisco Law School] and five MAs at Cal State University, East Bay, including in Anthropology, Mass Communications and Organizational/Consumer/ Audience Behavior, Popular Culture and Lifestyles, and Communication.

ABOUT AMERICAN LEADERSHIP BOOKS

American Leadership Books specializes in books on criminal justice, social issues, and politics and publishing a series of books on the subjects featured in these games. Many of these books were also turned into films.

The books include:

The Big Con: Scams Targeting Writers, the Victims, and How to Avoid Becoming a Victim,
I Was Scammed: The Many Scams Today and How to Avoid Becoming a Victim,
Scams in the Digital Age
Women with Partners in Prison
Women in Prison
The Costly U.S. Prison System
Crime in America
The Price of Justice in America

The films based on these books include:

Conned: A True Story (based on *The Big Con)*
Con Artists Unveiled (based on *I Was Scammed)*
Scams in the Digital Age (based on *Scams in the Digital Age)*
Love Behind Bars (based on *Women with Partners in Prison*)
Victims of Abuse (based on the forthcoming book *Abused)*

CONTACT INFORMATION

AMERICAN LEADERSHIP BOOKS, c/o
CHANGEMAKERS PUBLISHING
2415 San Ramon Valley Blvd., #4-366
San Ramon, CA 94583
925-804-6333
changemakerspub@att.net
www.changemakerspublishingandwriting.com